The Great Cataclysm.

2018

"The Great Cataclysm"

-The End of the Last Ice Age.

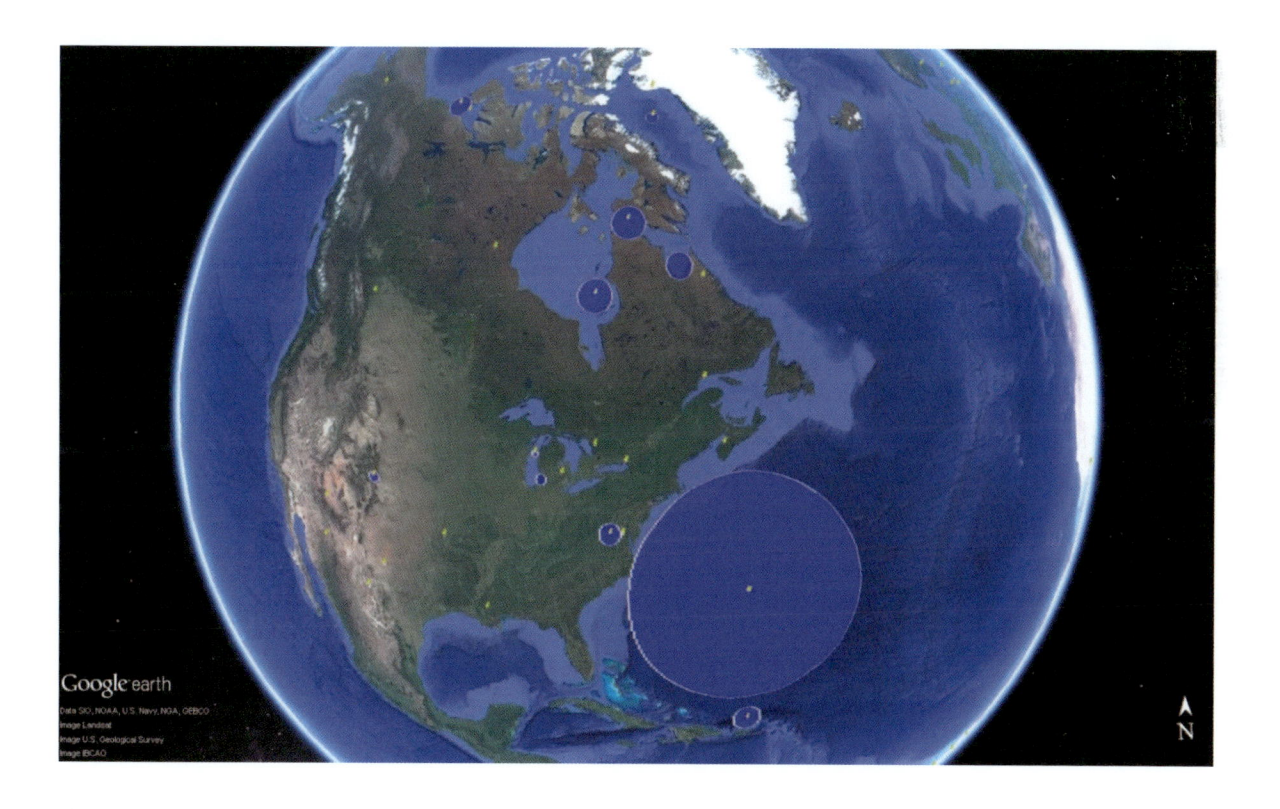

George Mitrovic

We are all led to believe that life on earth is a gradual and quiet process.
The Earth has experienced several warming and cooling episodes in its history. Simply put it was a period when the climate was much colder than it is today with much of the Earth's surface covered in ice. During the Ice Ages much of the Earth's surface was completely frozen and was barren tundra. Only a few plants, including evergreen trees, could grow in the frozen soil. The animals of this period were the woolly mammoth, the woolly rhinoceros, bears and reindeer. The traditional explanation is that the Ice Ages are caused by the Earth's position relative to the sun, its tilt and changes in the atmosphere. It is believed that small changes over a considerable period resulted in such dramatic changes in climate.

Is this the case though?

Are the changes almost unobtrusive and quietly approaching?

Some scientists state that we are currently living in an Ice Age called the Quaternary Ice Age and we are in a warmer stage of it known as the interglacial period when glaciers recede.

The Last Ice Age lasted over one hundred thousand years. It wasn't the first Ice Age and it will not be the last Ice Age that has happened on our planet. So when do we start to talk about the end of it? Let us start near the end. Around forty-five thousand years ago events began to occur that would cause changes to the Earth that would appear to be incomprehensible. These would include mass extinctions, the end of this particular Ice Age and changes to the face of the Earth and the seas completely. After this event this would be a totally different planet with totally different climates and life forms.

43rd to 30th Millenium BC.

Let us start with a meteoric impact event. They are quite rare aren't they.

The Cristie Impact Crater near the Kerguelen Islands in the Indian Ocean was created forty-five thousand years ago and is a meteoric impact crater fourteen kilometres in diameter. This is also called the Agatha Christie Crater. This is the first cometary or meteoric impact crater that we will meet.

What happens when we have an impact of this size? This impact would have propelled seismic waves across the face of the earth creating scores of earthquakes which themselves would have generated tsunamis hundreds of metres high. The sky would have become red hot as the atmosphere would fill with dust and at the same time the tops of the oceans would have boiled. The impact would have vapourized rocks which would have gone up into the atmosphere before condensing in to liquid droplets that solidified and fell back to the surface. This is as well as meteoric errata falling as well. This would be like the book of Revelations come to Earth. What was happening on Earth at this time. More than you may think as well.

Who were the witnesses to the destruction from the skies?

It is only one impact anyway. Who was around then to be affected by it?

Shanidar is a huge cave situated high above the Greater Zab River about 520 kilometres northeast of Abu Hureyra in the mountains of Kurdistan in Iraq. It has a long history. In this cave Ralph Solecki of the Smithsonian Institute found the remains of *Neanderthal man* who had lived here 45,000 years ago. According to classic evolutionism he ought to have evolved into modern man by then but he was still around. Several skeletons of *Neanderthals* were found who had died from a roof fall and had been buried ritualistically. Ashes and food remains over the graves hinted at a funeral feast and as well there were eight different types of pollen from wildflowers that suggested that the flowers covered the dead. Two of the skeletons were of an old man and a disabled woman indicating that the community looked after its frail and aged as well as having a possible religious belief in the afterlife. Ralph Solecki also discovered a slim almond shaped piece of copper with two equally spaced perforations at its end so that it could be worn as a pendant around the neck. The stratum it was found in was around 9,500 BC. Do we have artistic *Neanderthals*? Relics from Shanidar indicate that man had lived in the cave for one hundred thousand years from 100,000 BC. Some artifacts found on the floor were one hundred thousand years old. Also found at Shanidar which appeared to have been a summer encampment there have been found thousands of broken animal bones from immature wild sheep dating to 10,500 BC. This seems to show that the inhabitants practiced careful selection of wildstock. Were they still *Neanderthals* ? Or were there new residents at Shanidar?

We had witnesses to these meteoric events. Intelligent emotional people trying to survive in a harsh world.

What else was happening in this period? Around 42,000 BC world sea levels had dropped to sixty-one metres or two hundred feet below present sea levels. This means that glaciation had increased and more water was being locked up in glaciers. What this means is that the more ice is being created during a glacial period the lower the sea level as more water is held in the ice masses. You will find that sea levels go up and down during Ice Ages as well and quite dramatically.

Wolf bones found in Norway above the Arctic Circle indicate that the area must have had a temperate climate 42,000 years ago, not the grip of an Ice Age. When we allow for Earth tilts this makes sense. Where was the increasing glaciation though? Not in Norway. This is because Norway was nowhere near where it is now as the North Pole was in Hudson Bay in Canada. Yes, Hudson Bay in Canada was where the North Pole was. Before it moved again. What sort of momentous changes are these? What do you mean the North Pole has moved?

Geologists have discovered that the hard outer crust of our Earth floats on a molten mantle and that continents rest on separate tectonic plates. There are direct relationships between the sliding of the plates and the changing of the Polar Ice Caps. The Poles seem to stay stationary for about thirty thousand years and then shift for six thousand years and then stay put for another thirty thousand years. The last four rounds of the poles started one hundred and twenty thousand years ago when the North Pole installed itself in the Yukon Territory in Canada at 63 degrees north and 135 degrees west. Then it went to the Greenland Sea at 72 degrees north and 10 degrees east about 84,000 years ago. It then moved from 54,000 till 48,000 years ago and settled in the middle of the Hudson Bay area at 60 degrees north and 83 degrees west. It rested there for 30,000 years and then wandered from about 18,000 to about 12,000 years ago to its present location. Simultaneously the South Pole performed similar gyrations but in the opposite direction. The three moves prior to the last one were all in the Southern Indian Ocean and only the last one twelve thousand years ago ended up in Antarctica.

You will be learning a lot as you read this book. A lot that has not been taught but needs to be.

Now we head for South Africa. Early man was everywhere.

In the Ngwenya Hills in the Hhohho district of North-western Swaziland hematite, iron ore, was mined in 41,000 BC. The tunnels still exist. What did the residents need the iron for? In early times meteorites were the most common source of iron. This was different. This iron was being mined! Or should I say that the hematite was being mined and then who did the processing? And how?

Also around 41,000 BC in South Africa in Namibia in the Mountains of Fire strange discoveries were made on Mount Brandberg. During the 1952-1953 Thirstland Expeditions to southern and central Africa's desert zones, paintings were discovered that were quite unlike normal Bushman work. The central figure was a pretty white woman, young, graceful, her hair bobbed in the style of Ancient Egypt. She wore a beaded headdress, a garment resembling a modern jersey blouse, shorts, gloves, girdle and shoes similar to those worn in Mediterranean countries. Nearby there was a natural rock amphitheatre that appeared to have been used in ancient times. Behind her stands a thin man with a strange prickly rod in his hand and wearing a very complicated helmet with a kind of visor. Was he a kind of shaman? These are not cave people as we understand them.

Near here are more ancient iron ore mines discovered in South Africa that are 43,000 years old as well.

Were these early pictures of the Iron Mining Culture?

Were these the same people who were also mining iron ore in the Ngwenya Hills in Swaziland? Where is primitive man? This culture appears to be highly organized if it can create caves to mine iron ore and then possibly trade with it. Otherwise why dig it up?

World sea levels had risen by 40,000 BC and were now 42 metres or 138 feet lower than at present. This means that the glaciers were retreating as sealevels had risen nineteen metres in only one thousand years. This was a sudden rise of nineteen centimetres per year. How does this compare to modern sea level rises due to global warming.

Early estimates of twenty-first century sea level rises were around two metres in one hundred years. This was twenty centimetres per year. There is only one centimeter difference but no anthropogenic, or man made, cause.

Was this due to changes in glaciation? It can't be anthropogenic global warming as carbon-producing industry was not particulary abundant. We haven't found the smelters for the Ngenya Hills iron mines yet? There must have been a lot of them if they influenced world climatic levels to this extent.

Polished stone axes have been found in Niugini which is a large island north of Australia dating back to 40,000 BC. Had these people managed to cross the Wallis Line which has always been open water and never above sea level? Apparently the Australian Aboriginals had done this 74,000 years before. The Wallis line always has been sea and never land. Massive cross currents between two continental shelfs make it impossible for anything to drift across it. Only pilotable watercraft could do it. How did the Aborigines do this? How did the Nuiginians do it? They are not maritime cultures but they must have been when they were on their way to Australia and Niugini. The Wallis Line separates the world of the marsupials from the southeast Asian non-marsupials. Nothing can drift across it due to opposing currents. In this period there were two large land masses on either side of the Wallis Line. To the west was Sundaland which included Indonesia, Bali, Borneo and most of South East Asia. To the east was Sahul which included the larger land masses of Australia and Niugini and the islands of Timor and Lombok amongst others. Nothing could ever drift between them. The human race was everywhere and much more advanced than is commonly known. But it still needed navigable and guidable boats to get across the Wallis Line. It could not drift across.

Early man seemed to be either mining or refining iron ore. In 1996 a team of scientists from the Roviri I Virgili University in Tarragonna in Spain unearthed fifteen furnaces at Capellades. Each of these furnaces, apparently built by *Neanderthals*, served a different function according to its size. Some were ovens, some were hearths and some were blast furnaces. There was also an astonishing variety of stone and bone tools and many traces of wooden utensils. Iron has been mined in this area since the dawn of time especially along the Rio Tinto. This was around 40,000 BC. If iron was being mined and worked in numerous places across the earth why did it not accelerate the progression of civilization in this period? Or had it? Or are there no relics to be found? Who were the Cavemen and Cavewomen? Or did we hit a gigantic full stop further along the way?

Do civilizations come and go? Had the iron relics that were created by these early cultures merely rusted away?

The coastal areas of the world in this period are now below sea level and coastal areas are generally where civilizations lie. Iron does not survive in the sea for very long, let alone forty thousand years. We might have to look out on our continental shelves for early sites as that is where the trade centres would be as they generally are now.

Now we are in Russia. As I said, humans were everywhere.

In 1964, 125 miles northeast of Moscow near Valdimir on the Sungir River, Otto Bader discovered burial places that from the strata that they were found in must have been at least 30,000 to 40,000 years old. The bodies were lying in frozen ground and were in good condition and beside them were 7,500 bone ornaments and carved beads that had been sewn into the garments. The clothing consisted of trousers, pullovers and leather shoes. These people were not *Neanderthals*.

Where are the skin-clad *Neanderthals*?

Also in this same area the skeleton of a man of about 50 years of age, with wide shoulders and a height of 175 centimeters and with unusually long shin bones like a sprinter was found. Not Ape-like at all and with a very large brain, high forehead, and a pronounced modern chin. He lived 52,000 years ago by hunting mammoth and reindeer. The man wore a large pair of trousers and a very practical jacket with ivory badges. The man also wore a pair of hide shoes. Giant hooked mammoth tusks that had been straightened out to make spears were also found here. They are 27,000 years old. In 1964 the bodies of two boys from 27,000 BC were exhumed and were found to be wearing shirts, coats, leather trousers and hats. A bone needle that would be a replica of our steel needles was found here dating back 27,000 years around 25,000 BC.

We have continuous occupation and remarkable fashion achievements. Why did this civilization not continue forwards? Or had it?

How sure are we with our dating systems? These are generally based on radiation decay over time. Over a considerable stable time where everything decays at a steady rate. What if this is not the case? What if everything is topsy turvy?

A human habitation site was discovered in Alaska in Yukon-Koyukuk on the confluence of the Old Crow River and Porcupine Rivers. Remains have been found here dating back 20,000 to 40,000 years though other scientists state that this actually goes back 68,000 years. As usual they can't agree. Bone fragments suggest that man lived here around 40,000 BC. The processed animal bones indicate that the humans in this area hunted mammoth, giant beavers, ground sloths, camels, several kinds of horses, giant bison, short-faced bears, American lions and many other extinct animals. If the climate here was snowy and below freezing how did anything grow for these giant animals to feed on?, These animals numbered in their millions, This is the same problem with the Siberian woolly mammoths. How did their giant bulks get enough nutrition if they lived in the perpetual snow? And where did the buttercups grow that were found in their mouths? Along with the branches and tree leaves? Early man was not supposed to have arrived in North America until 15,500 years ago or 13,500 BC. Alaska contrary to common opinion was never covered in glaciation.

Human life on earth seemed to be just bubbling along quite nicely.

Suddenly something totally unpredicted occurred that would signal the changes that would create havoc as life, climate and geology changed on the planet.

Around 39,000 BC there was a wavering of the Earth's magnetic field called an anomaly or excursion. In this case Earth's magnetic field wavered and almost reversed meaning that the Earth's magnetic pole flipped down and almost became the South Pole. Radiocarbon levels increase in these anomalies or excursions. How much do these reversals and radiocarbon increases effect our dating systems? There was a massive and sudden peak in radiocarbon in the atmosphere of our planet. This could be caused by a supernova or explosion of a massive star when there is a pulse of radiation hitting the Earth's atmosphere that creates a surge of radiocarbon. This can also be caused by a solar flare. A third probable cause could be a declining of the Earth's magnetic field or a thinning of the atmosphere. Another cause can be comets or asteroids impacting with the Earth. Or all of the above?

At the same time there were massive extinctions of millions of animals in Australia and Southeast Asia. Many animals in Australia larger than two hundred pounds became extinct including the giant wombat and the marsupial lion amongst others. All in all forty-four species became extinct. Mind you other paleontologist state that the great Australian extinction occurred around nine to ten thousand BC. Once again the paleontologists and geologists are still arguing.

One theory is that the light and radiation from a massive Supernova explosion had arrived on Earth and hit the side facing it at the time, namely Australia and Southeast Asia. Within seconds the radiation filled the skies with light as it hit the atmosphere and damaged oxygen in the air to create ozone. The blast probably lasted only ten seconds but it killed almost every living thing that had been exposed to it. As vegetation wilted and died the large megafauna in these regions became extinct as there was too little food for the survivors to survive. The lightning-filled unstable skies set of mega firestorms that also destroyed food and created nuclear winter-like weather anomalies including loss of sunlight for plant growth. Where were the Aboriginal inhabitants of Australia at this time? How did they survive this radiation storm?

Another anomaly mentioned by Richard Firestone, Allen West and Simon Warwick-Smith is that in this period *Neanderthal man* started disappearing and *Cro-Magnon* man suddenly mutated into modern man. The wolf metamorphosed into the dog in this period as well. They state that there must have been a mutation in man's brain as well in this period with a flourishing of art and music, religious practices and more sophisticated toolmaking techniques. Was this mutation caused by the radiation storm?

Richard Firestone, Allen West and Simon Warwick-Smith also suggest that there was a split into the two races of Asian and Caucasian in this period due to a major mutation. There is evidence that from this time every paleo-Indian race, Asian, descended from the one small group of people from a previously larger population that had perished. This is technically called a bottleneck. The human population of Earth had mysteriously dropped from perhaps a million people to several thousand. Was this all due to radiation from a supernova that actually only lasted a few seconds or from the consequences of a nuclear winter? And how many times has this happened in the past? Humans had almost become extinct and possibly not for the first time.

In the Miri Forest in the Gunung Subis Mountains on what is now the west coast of Sarawak in Borneo a network of caves was found that had been hollowed out on a cathedral-like scale. Among these colossal caves there are fabrics of extreme finesse and delicacy. Who made the fabrics or the cave system is a total mystery. Many of these caves have perfectly formed ninety degree angles, flat walls and ceilings and stretch for miles. These are the Niah Caves and this was originally in Sundaland. Remains in these caves date back 40,000 years to 38,000 BC. More on the now sunken land mass of Sundaland later.

More caves? Were the inhabitants hiding from stellar radiation or some disaster coming from the skies?

Lake Ladoga in Karelia in Russia is the site of an astrobleme or Star Wound created forty thousand years ago. The crater is in the northern part of this lake which is the largest in Europe. This is evidenced by the finding of impact breccia and shatter cones. The crater is eighty kilometres across. Other sources state that this occurred sixty-five thousand years ago. Twenty-seven thousand years is nothing. This is a massive meteor crater. We have also read how much damage would have been caused by this impact.

A fire pit found on Santa Rosa Island in the Channel Islands in Santa Barbara County in California showed no traces of Carbon-14 at all thus showing a minimum age of 40,000 years or 38,000 BC. There were crude chopping tools in the pit along with the bones of a full-sized species of mammoth. Amongst the 40,000 year old remains were man-made chopper tools and dwarf mammoth bones. Carbon 14 or radiocarbon dating generally can only go back fifty thousand years. So were these fire pits even older?

Once again when did man first show up in the Americas?

The remains of a giant man with double rows of dentition, or double dentition, were found on Santa Barbara Island in the Nineteenth Century. He and his descendants feasted on the small elephants that used to live here and vanished countless ages ago. The giant with the double teeth was similar to the skeleton found at Lompock Rancho on the nearby mainland in 1833. The giants of these islands were the extinct Chumash. The giant human had a double row of upper teeth and a double row of lower teeth. Double dentition is not unknown.

There are the remains of a human settlement on Santa Barbara Island that dating by C-14 dating processes goes back 29,600 years. How popular has California been over the millennia?

Lime Creek Man left tools of bone and stone dating back 40,000 years from the third Interglacial period. Lime Creek is near Stockville in Frontier County in Nebraska.

The skull of a *Neanderthal man* was found in Broken Hill, now Kabwe, in what was formerly Northern Rhodesia but is now known as Zambia. Clearly demonstrated on one side of the skull is a hole made by a bullet and on the other side the aperture where the bullet went out again and where the skull was completely shattered. This skull though is 40,000 years old and there are no radial cracks which are usually present if the injury was caused by a tusk, horn or cold weapon such as an arrow or spear.

Interestingly enough manganese was also mined here 28,000 years ago and this was verified by the carbon dating of charcoal on the site.

Once again the ancient predeliction for mining by ancient man.

In 1921 the skeleton of an early man called "Rhodesia Man" was found here as well dating back to this same period.

Near China Lake in Kern County in California there are drawings that show rings, stars, spheres, many coloured rays and figures of Gods. These are dated to 37,000 BC. Oddly enough on one rock a drawing apparently of a Tyrannosaurus Rex with something in its front paws resembling a man was found. An old picture amongst later ones? A fraud in the middle of nowhere? More dinosaurs? A bit late for them isn't it? China Lake is a human habitation site. How old is actually the mystery. The area was not desert in this period.

And the archaeology keeps apparently getting weirder. Man was everywhere.

In 1958 near Lewisville or Louisville in Denton County in Texas stone tools and burned animal bones were found in association with hearths. These were radiocarbon dated as 38,000 years old.

From 1932 onwards projectiles similar to the Folsom ones were found on the border between Texas and New Mexico and west to Naco in Arizona that are 38,000 years old. The Folsom tools are from ten to twelve centimetres long, 4.8 inches, and date back at least to 8,000 BC to 11,000 BC. They were able to penetrate the skull bone of a mammoth. These tools are 27,000 years older.

The period of 35,000 BC was an interglacial period when temperatures were much higher than present.

Around 33,000 BC world sea levels were sixty metres or 180 feet lower than the present. In the previous eleven thousand years the levels had risen only twenty feet. This indicated that the ice masses and glaciers were still growing as more water was being frozen.

Paintings, engravings and drawings of quite a mature and sophisticated style have been found in the Chauvet Caves in the Ardenne of southern France. There are at least 420 animal figures including many now extinct species.

Early man inhabited the Grimaldi Caves, three miles west of Ventimiglia in Liguria in Italy, as early as 35,000 years ago. The *Cro-Magnon* skeletons were buried carefully, their bodies having traces of red ochre sprinkled over them. They were still wearing ornaments in the form of necklaces and bracelets of shells and animal teeth. They were of a man, a woman and an adolescent child. There were also two fossilized skeletons found here as well. They are of an old woman and a young boy. They have been buried together with their heads protected by a horizontal slab placed on two vertical stones. The shapes of the skulls and the facial features as well as the length of the forearms indicate that they are Negroid. Negroes were not supposed to have lived in Europe in the Paleolithic period. In this period the Mediterranean Sea was still a series of landlocked lakes in a huge valley so crossing over was not a great problem. Humans had been living in the Mediterranean area for many millennia.

At the site of Dyuktai in the Aldan River Valley in Khabarovsk in Siberia three hundred miles west from Okhotsk the remains of humans have been found whose culture was very similar to humans in North America at the same time. These remains date back to 35,000 years. The Dyuktai people were hunters and many bones of horses and mammoth were found here. The remains were discovered by Yuri Mochanov.

So we can see that much of the earth was inhabited by viable cultures of intelligent creative people.

Around 32,000 BC strange things occurred that would eventually lead to the end of the Last Ice Age. There was another wavering of the Earth's magnetic field. Another anomaly or excursion. In this case Earth's magnetic field wavered and almost reversed meaning that the Earth's magnetic pole flipped down and almost became the South Pole. The previous anomaly was only seven thousand years before. Would there be another bottleneck?

In the period of 32,000 BC there was a massive peak in radiocarbon in the atmosphere of our planet. This as I have explained before could have been caused by a supernova or explosion of a massive star when there is a pulse of radiation hitting the Earth's atmosphere that creates a surge of radiocarbon. Or it could have been caused by a solar flare. A third probable cause could be a declining of the Earth's magnetic field or a thinning of the atmosphere. Another cause can be comets or asteroids impacting with the Earth. Hadn't we just had one of these only seven thousand years before? Mind you we need to remember the the general accuracy of geological dating, or lack of accuracy. A massive radiation storm would certainly upset radiocarbon dating which is based on a steady release of radiation. What would enormous radiation surges do to the results?

Richard Firestone, Allen West and Simon Warwick-Smith state that in this same period the second blast wave from a theorized exploding supernova was hitting the Earth. This was not as obvious as the first blast seven thousand years earlier. There would have been a profusion of shooting stars as particles hit the upper atmosphere and more damage to the ozone layer as radiation levels edged upwards. There were intermittent strikes on earth of larger particles that exploded on impact showering everything around with high velocity micrometeoric material which was fatal. These incoming grains were high in toxic metals and radiation and led to massive extinctions. This time the radiation storm hit the Northern Hemisphere.

Does this sound familiar? The same event had occurred in 39,000 BC but in the Southern Hemisphere. A different face of the Earth was facing the interstellar attacker this time.

30th to 20th Millenium BC.

Somehow humans survived the radiation storms. In the Southern Hemisphere.

The Boquierao do Sitio da Pedra Furada also called Toca do Boquierao da Pedra Furada in Piaui State in northeastern Brazil is a rock shelter inhabited by early man. Archaeologists dug through a three-metre deposit of sediment that was found to contain human occupational debris at all levels. At the lowest levels were big circular hearths as well as pebble tools, denticulates, burins, retouched flakes and double-edged flakes. There were also painted fragments of rock spalled or broken from the cave walls that suggest rock painting. The age was around 32,000 years or 30,000 BC. The older layer of painted sandstone has two red lines painted on it and are possibly part of a figure. Pebble tools were also found here as well as hearths, one of which was dated at 32,000 years old. Some estimates are that the site dates back to 50,000 BC. Take your pick. Three fossilized teeth as well as part of a human skull have been found here that are fifteen thousand years old. Pedra Furada means perforated rock and there are over four hundred archeological sites here. It was a very popular area though man was not supposed to exist in this part of the Americas at this time. Somebody forgot to tell him or them though.

Another anomalous site is at Monte Verde on Chinchihuapi Creek northwest of Puerto Montt in Chile. Originally comprising two sites a few hundred feet apart that had been flooded under peat and then preserved. Hundreds of stone, bone and wooden tools, animal bones, even mastodon bones and flesh, cylindrical spear points, timber poles with attached pieces of knotted reeds that were the remains of hide covered huts, as well as remains of firepits and a childs footprint were found in undisturbed layers so finding relationships and sequences was simple to establish. The site is located in a boggy area in which perishable plant and animal matter was well preserved. Two pebble tools were found hafted to wooden planks. Twelve architectural foundations made of cut wooden planks and small tree trunks staked in place were also found, as were large communal hearths as well as small charcoal ovens lined with clay. Some of the stored clay bore the footprint of a child 8 to 10 years old. Three crude wooden mortars were found, held in place by wooden stakes. Grinding stones were uncovered along with the remains of wild potatoes, medicinal plants and seacoast plants with a high salt content. There were the remains of domestic amenities. At the oldest level there was found a split basalt pebble which is a kind of primitive tool, some wood fragments, two modified stones and some charcoal dated at 33,000 years old, 31,000 BC.

On Rancho La Amapoloa 1.5 kilometres southeast of El Cedral in San Luis Potosi State in Mexico at an altitude of 1,700 metres human artifacts as well as elephant remains and indications of the use of fire have been found in undisturbed strata up to thirty-three thousand years old or 31,000 BC. A hearth here was found surrounded by elephant bones. Roasted elephant was also popular in what are now the Channel Islands in California.

There were some survivors in the Northern Hemisphere as well.

In 1823 a modern human skeleton lacking a head was found buried deep in ancient strata at Paviland in Wales. The skeleton had been buried for so long that it had been stained by the earth and was called the "Red Lady of Paviland". In fact it was actually a man. And it had been dyed with red ochre. The body, discovered in Goat's Hole Cave on the Gower Peninsula, is 33,000 years old. Red ochre or iron ore was used in many cultures worldwide to paint people for ceremonies and to paint the deceased. Was this what the early iron ore mines were for? So why the smelters in some of them?

Around 30,000 BC world sea levels are 52 meters or 170 feet below present levels. The sea levels are rising again so glaciation is increasing again. As you can see rising and lowering sea levels are not rare or uncommon you just have to look at the larger vision.

Russian Scientists concluded that the Arctic Ocean was warm during most of the last Ice Age and the period 32,000 to 18,000 years ago as being particularly warm. What do they mean by warm? Warmer than now or actually warm. So the Arctic Ocean was not covered with two miles of ice? In this period the North Pole was where Hudson Bay is now.

The skull of a child was found in Taber, Alberta, in 1961 that was thirty-two thousand years old. There were fire pits found here from the same period. Where is the two mile thick ice mass that is supposed to be here? It was further south.

The Elegante Crater near Sonoyta in the Sonora Desert in Mexico is 1.6 kilometres in diameter from rim crest to rim crest and thirty-two thousand years old. The crater is circular and has a peak in the middle. The crater is surrounded by a blanket of tuff breccia. To the northeast is another crater almost the same size as well as fragments of numerous other ones. The mountain that the craters are on is on the shore of the Gulf of California.

This is only a small cluster anyway.

The Krugloye Ozero in Novosibirsk in Russia is an impact crater that is six hundred metres in diameter and thirty thousand years old. Just a coincidence that they are from the same period?

These are only two craters after all in one thousand years.

The remains of a race of people called the Boskop have been found in the western coastal areas of South Africa around Potchefstroon. The Boskop were an unknown branch of humanity that rivalled in brain volume people in Europe, ancient or modern. The ratio of cranium to face is five to one whereas in Europeans it is three to one. The teeth were also very modern and not suitable for a primitive hunter. The teeth were free of any dental ills. One indication of the intelligence of a hominoid species is the degree of erectness of their stance which influences how much blood can flow to the brain therefore stimulating brain growth and theoretically the development of a species. The more blood flowing to the brain, the more the brain can develop and the larger the cranial capacity. This should only come about after an incredibly long period of evolution. Following this logic then if there were a race of hominoids more advanced than we are then their spines would join at the base of the cranium making them more erect than we are with our craniums joining further back. The Boskop had an average cranial capacity of 1800 centimetres compared to ours which average out at 1300 centimetres. The Boskop skeletons are also splendidly evolved. The Boskop though died 30,000 years ago and were only around for a very short period. Were they visiting? No remains of a culture have been found with them. Only the skeletons with the odd skulls. Were they visiting and from where?

World sea levels around 29,000 BC were now 48 metres or 157 feet below sea level. In one thousand years they had risen four metres. Glaciation had increased again.

Four kilometres north of Hueyatlaco in Mexico a bone sample of a human found here with mollusk shells indicates an age of 21,850 years to 30,600 years, 19,850 BC to 28,600 BC, using the carbon dating method. This was near Caulapan on the Tetela Peninsula in Puebla in Mexico.

Around 28,000 BC world sea levels were now 96 metres lower than one thousand years before at 134 metres below present levels indicating massive increases in glaciation. This indicates a phenomenal drop in sealevels of 48 metres over one thousand years! This was a drop of half a metre per year indicating a massive increase in glaciation. This makes modern sea level increases seem inconsequential.

Around 30,000 years ago the world plunged into a full glacial cycle with the global climate turning colder, vegetation thinning out that allowed steppes to take over many countries. Lush vegetation was limited to springs at lower altitudes. The ice caps grew larger and the sea levels dropped by four hundred feet increasing the size of the coastal plains. Remember though that the North Pole is not in the location that it is in now so the centre of the ice mass is in a different location and radiates out from this different centre. It was not an icemass coming from where the North Pole is now but from where it was at the time in the Hudson Bay area. Increases in albedo increased the ice cover in this period. Albedo is the measure of the diffuse reflection of solar radiation out of the total solar radiation received by an astronomical body. In simple terms albedo is the ratio of light reflected by a planet to that received by it. The increased ice masses were reflecting heat and light making the planet colder.

Glacial periods are periods of cooler and drier climates over most of the Earth with large land and sea ice masses extending outwards from the Poles. The snow line is lower so mountain glaciers occur in unglaciated areas and at lower altitudes. Sea levels also drop due to accumulaton of ice in the ice caps. Then ocean circulation patterns are changed. Whilst this is occurring positive feedback processes occur wherein the albedo of the Earth is influenced. Snow albedo is where local snowfalls reflect away sunlight and if no temperature increases occur increase in mass causing more snowfall until a natural feedback mechanism such as a temperature rise causes it to reverse. Ice and snow reflect back more of the sun's energy and absorb less. The air temperature decreases and ice and snowfields grow until competition with negative feedback mechanisms causes equilibrium and warming begins again. The reduction in forests also increases albedo.

Around 28,999 BC the Adriatic Sea between Italy and the Balkan states of Europe was dry land composed of three hundred miles of coastal plain. The sea as being swallowed up by the glaciers.

The monolith of Itaquatiara de Inga is covered in bas-relief inscriptions that resemble those of the Near East. The consensus of opinion is that the monolith was engraved twenty to thirty thousand years ago, 8,000 BC to 28,000 BC. Carved by whom and why? After all writing was not supposed to be invented until the arrival of the Sumerians around 3,000 BC. Mind you we will meet a few other cultures that predate this age as well. Itaquatiara de Inga is in Bahia State in Brazil. Did civilization travel from South America to Europe and North America after the two radiation storms and the population bottleneck.

A human skull found at Otavalo in Imbabura in Ecuador has been dated at 30,000 years or 28,000 BC, and possibly older.

There are impressionistic cave portraits of human individuals at Grotte de la Marche at Lussac-les-Chateaux in Vienne in Poitou-Charente and they are of clean-shaven people with no beards or trimmed beards and hair and wearing shaped clothing rather than bearskins. These portraits are 30,000 years old.

In August 1969 the completely preserved bodies of two men that were 30,000 years old from 28,000 BC were found in the Cave of the King, a grotto near the Bay of Biscay, eighteen kilometres from Santander in Cantabria in Spain. The bodies, which had been preserved completely by porous clay, seemed to be identical to modern man. We are still missing a lot of the *Neanderthals*. Where had they gone?

Tiny figurines carved out of mammoth ivory have been found at Hohle Fels in the Swabian Jura in Swabia in Baden-Wurrtemberg in Germany dating back thirty thousand years. The carvings depict a very naturalistic duck-like bird, an animal that resembles a horse and a half human half animal creature that appears to have the body of a man but the face of a lion. This is not the primitive art supposed to have been created in this period. None of the carvings are longer than five centimetres and are incredibly detailed. Nearby at Vogleherd, Geissenkloesterle and Hohlenstein-Stadel, seventeen other minute and delicate sculptures from the same period, the Aurignacian Period, have also been discovered. One of these carvings was a musical pipe or primitive flute made of swan bone and all were found in the Ach Valley and Lone Valley southwest of Ulm. The statuette of the lion-headed man was found at Hohlenstein and was in strata 32,000 years old around 30,000 BC. Years later museum officials were presented with a beautifully carved ivory muzzle that fitted the lion head perfectly. If we had ivory muzzles in Southern Germany then it is not difficult to believe in horse bridles in La Marche in France.

Around 28,000 BC the Argive Plain in Greece stretched six miles further south in Greece due to sudden lowering of sea levels.

On the slopes of the volcano Mont Musine in Turin in Italy there are 30,000 year old carvings of strange disc shaped objects. The local legend around Mount Musine is that it is the home of the Chariot of Herod which comes out of the mountain with fire and blazing lights. As recently as the 1960s the Chariot was still being reported to be seen.

In 1933 during excavations Dr. F. Weidenreich discovered a number of skulls and skeletons at Chou-kou-tien now called Zhoukoudian near Beijing in China. One skull belonged to an old European, another to a woman with a narrow head, typically Melanesian in character, and a distinctively Eskimo young woman. These skulls are 30,000 years old from 28,000 BC. The three skulls were underneath a great layer of thousands of animal bones dating from the cataclysm of 9,000-10,000 BC. The caves are crammed with assorted human and animal bones in astonishing diversity, apparently swept there from far away by a titanic flood, or judging by the age differences, by several different floods. In these caves near Peking, Beijing, in China the bones of mammoths and buffalos have been found in association with human remains from the end of the last Ice Age. The remains of Peking Man were also found here in the 1920s. Peking Man dates back around 750,000 years ago. The ancients liked long settled cave systems.

At Numazu in Shizuoka Prefecture near Tokyo a four metre deep Stone Age excavation revealed the remains of stone ovens indicating a high degree of culinary sophistication. There were many left over elephant bones here as well. The use of ovens helped cook the meat more delicately without charring it. These ovens date back to 28,000 BC. Elephant is still the meat of choice in ancient times.

Mount Oikeyama is in the Shirabiso Highland in Nagano Prefecture in Japan. The Mount Oikeyama Impact Structure in Nagano Prefecture in Japan is nine hundred metres in diameter and thirty thousand years old. It would have certainly had an impact on the nearby Numazu oven culture. Mount Oikeyama is only fifty miles from Numazu.

Now for an anomalous nineteenth century report. It helps break the boredom of reading about more impact events than you ever thought possible. At a depth of seventy feet below ground level the remains of baked brick pavements and cisterns have been reported to have been found near Memphis, Tennessee in the Nineteenth Century. This was in Shelby County. The bricks resembled Roman bricks. There were also interlocking modular ceramic drainage pipes of modern appearance. The strata that they were in dated back to 28,000 BC and was under glacial strata from the last Ice Age. Is our dating still out or do we have to more closely look at the revolving door of civilization?

In the Musuem of Paleontology in Moscow there is a skeleton of a bison whose brow has been pierced by a single neat shot. This was in 28,000 BC when weapons were still made by flaking stones and the most modern weapon was the stone axe which was incapable of shattering the skull. The piercing is the same as that produced by a bullet! No other weapon could have made it. But in twenty-eight thousand BC? The bison was found the Yakutsk Republic. The fossil was found to the west of the River Lena and the edge of the aperture or hole was calcified indicating that the animal survived its nasty encounter. This is another Russian fossil shooting. Other opinions state that the bison was shot hundreds of thousands of years ago. Either way who was shooting bison with guns in the remote past well before the invention of the firearm?

Or was it a micrometeorite?

There are the remains of a human settlement on Santa Rosa Island in Santa Barbara County in California that dates back by C-14 dating to 29,600 years, 27,600 BC. There seemed to be a civilization living here on the Channel Islands off the California coast.

Five miles out in the Mediterranean Sea off Marseilles in the Provence-Alpes-Cote-d'Azur in France Jacques Mayol has explored a mile long shoal running at a depth of 60 to 120 feet with vertical shafts, quarries and slagheaps lying outside the shafts. This was a man-worked mining complex contemporary with the *Cro-Magnon* period. There are also horizontal tunnels and smelting facilities from a time before the Mediterranean Sea was flooded. The discovery off Marseille was a man-worked iron ore mining and refining complex from before the Great Cataclysm due to its position on the sea shelf which was only above sealevel prior to the cataclysm many thousands of years before the accepted invention of iron.

Also in the area of Marseille is an undisturbed cave called Cosquer Cave which is 137 feet under the present sea surface which was discovered by divers in 1991. The cave dates from 27,000 BC to 18,000 BC. There are Magdalenian cave paintings here indicating that the site was once above sea level. Allowing for the three hundred foot rise in sea levels at the beginning of the Holocene Period there would have been a large continental shelf below here. Was this a sacred site for the iron workers? Our major centres are coastal with easy access to trade. In the remote past water traffic was the most common form of transport between different trading or mining areas. Rivers, seas and lakes were the ancients' super highways.

A group of five dwellings have been found near Dolni Vestonice in the Czech Republic dating back 28,000 years or 26,000 BC. The largest was over fifty feet long. Nearby were the remains of a large pottery kiln that was used apparently only for firing small clay figurines as no domestic pottery has been found here. Over 10,000 pieces of fired clay were found here. This would be more than for domestic use and flints found here indicate that it was a trading area over a large area. Mammoth ivory carvings have also been found here. There were also small bone carvings of the heads of wolves and bears and even rhinoceroses. Around the community was a fence and there was a substantial pile of mammoth bones in a marsh.

Another feature of Dolni Vestonice are the houses built over pits allegedly chiseled from the frozen peat with hand axes. The tentlike structures were arched with poles which were draped with animal skins stitched together with animal sinews. These hides were anchored to the ground by the hipbones of enormous animals and the skulls of reindeer, mammoths and woolly rhinoceroses. The tusks themselves were used as fuel for hearths. Was the peat even frozen though or is this a modern assumption?

A mammoth tusk found here has an engraved map of the Dolni Vestonice region with the Dyja River represented as well as the ridges and erosion slopes of the loess plain. At the centre of the map is a double motif of a complete circle and a half circle that might mark a significant site to the creators of the map. This would make it one of the oldest maps in the world if not the oldest.

25th to 21st Millenium BC.

A settlement has been excavated at Predmosti in North Moravia in the Czech Republic where the remains of mammoths and men were found together. The skeletons of 800 to 1,000 mammoths were found. The shoulder blades of mammoths were used in the construction of human graves. The remains date back to 25,000 BC.

Malta is sixty miles northwest of Lake Baikal in Siberia on the Angara River near Irkutsk. Indications of human presence here have been found dating back 26,000 years or 24,000 BC. Finds included bone needles that indicated that sewn clothes were being used. Sewn clothes have been found in other sites in this period such as in Vladimir in Russia.

Along the Azores-Gibraltar ridge numerous remains of elephants have been found in at least forty different sites. Some of these sites were at depths of only 360 feet below sealevel. Elephant tusks were found in submerged shorelines, sandbanks and shorelines that were once above sealevel as well as other places that were all originally above sealevel as recently as 25,000 years ago and possibly much more recently. The elephants weren't just going out for an oceanic swim were they?

Incidentally Plato mentions elephants on the lost island of Atlantis. Another coincidence? Or actual memories? Allowing for a three hundred and sixty foot drop in sealevels the continental shelf around the Azores would be quite large. Or was Plato remembering mammoths and mastodons?

Hematite iron mines in Malawi in Africa were worked with rich veins of copper extracted as well dating back 25,000 years. Why mine for copper unless you can process it. The same goes for iron ore. Where was this processing being done? Possibly on the now vanished coast lines?

In 1952 Dr Paul Sears of Yale University dug up some maize pollen grains from 240 feet below the bed of Lake Texcoco, the dried up lake that Mexico City is built on. Maize is the most highly developed agricultural plant in the world. The pollen grains were 25,000 years old. This indicates that someone was harvesting domesticated maize twenty thousand years before the accepted dates of crop harvesting. This also indicates the presence of established communities capable of staying in the one area so as to develop, tend and harvest crops. Semitic style figurines have been found in the area. They have also been found on the Ecuadorian coast and also in San Miguel de Allende, Mexico. They are believed to be much later than the maize but still a mystery. How old are these? Farming was only supposed to start around 9,000 BC in the Levant or Middle East.

Lake Bonneville which was a massive glacial lake suddenly crested and covered over twenty thousand miles where Utah, Nevada and Idaho join together. The Great Salt Lakes are where this massive freshwater lake used to be. What caused the massive cresting or peaking event 25,000 years ago?

In 1976 a mammoth bone spear point was found in Bluefish Cave near the Bluefish River about fifty miles southwest of Old Crow in the Yukon Territories in Canada. Human worked mammoth flakes have also been found here as well. There are three caves at the site and one of them contains animal bones that appear to have tool marks on them.

A carved bone scraper was found at Old Crow River-Porcupine River in Yukon-Koyukuk in Alaska a few miles from the Alaskan border that was radiocarbon dated at between 25,000 to 32,000 years old or 23,000 BC to 32,000 BC. A human habitation site was discovered here also dating back 20,000 to 40,000 years. Alaska was never glaciated in the last Ice Age.

Flint points found in Sandia Cave in Las Huertas Canyon in Bernalillo County in New Mexico were dated to 25,000 years old. Sandia man at the latest was making leaf shaped spearheads during the last retreat of the glaciers 25,000 years ago around 23,000 BC. The remains of primitive Horses were also found along with those of early man. Sandia Cave is in Las Huertos Canyon in the Sandia Mountains. Implements of advanced type, Folsom Points, were discovered beneath a layer of stalagmites that were 25,000 years old. Folsom blades were found embedded in travertine a quarter of a million years old. Fire hearth and leaf shaped points have also been found. The artifacts are cemented in the crust under the stalagmites. How old are the Folsom Points though? Once again the disparity in geological dates.

Haematite iron mines in Zambia were worked with rich veins of copper extracted dating back 25,000 years, 23,000 BC. Why this prehistoric iron and copper mining? Unfortunately whatever they made with the iron or copper would have rusted away by now.

So far we have iron mines in South Africa, Malawi and now Zambia as well as in the south of France. We also have had iron ore smelting near Marseille on the continental shelf in the Mediterranean Sea. This would be where other smelting works would have also been.

Lake Constance or the Bodensee on the border of Germany and Switzerland is an impact crater around twenty-four thousand years old. The crater is twenty kilometres in diameter.

In gravel pits near Edmonton in Alberta in Canada are the remains of extinct giant bears, American lions, mastodons, bison, ground sloth, camels and horses. Where did these animals come from if this area was supposed to be covered by a glacial layer hundreds of metres thick in this period? What did they eat? They weren't fasting. This was in 21,000 BC.

At the site of Tlapacoya near Ixtapaluca in Anahuac State in Mexico the traces of human beings have been found that date back 23,000 years to 21,000 BC. Charcoal and blades have also been found here. Did this culture cultivate and develop maize in this remote period that was found nearby in Lake Texcoco?

There are inscriptions at Wari, or Huari, near Quinua in Ayacucho in Peru that resemble shorthand. There is also a depiction of a large running bird and a horse's head. The civilization that wrote the inscriptions and made the drawings is unknown. Ayacucho is two hundred miles southeast of Lima. These petroglyphs are believed to be 23,000 years old. Stone tools found here are at least twenty thousand to twenty-three thousand years old. Within caves in the area remains were found of hunting people who by 5,000 BC, very similar to Tehuacan in Mexico, were cultivating crops as well as domesticating the guinea pig and the llama. The gourd and squash were first grown followed by cotton, earlier than in Mexico. The Peruvian coast in this period changed from that of fog meadows that were fed by seamist that provided pasture and moisture for plant life to a very arid one where plants and animals began to vanish.

20 th Millenium BC.

xxx

Now for an anomalous archeological find. In Plateau Valley in Mesa County in Colorado Tom Kenny, a farmer, in 1936, found at a depth of three metres under the ground a piece of pavement consisting of handmade, smooth, symmetrical tiles. The analysis of the mortar showed that it was of different material than that available in the valley. Experts stated that it dated from 20,000 to 80,000 years old. Plateau Valley is on the western slopes of the Rocky Mountains. Other references state that the tiled pavement was 12,000,000 to 26,000,000 years old. The tiles measured 127 millimetres square with unknown separating mortar. Why all the weird dates? Even twenty thousand years old is amazing for North America.

Other sources state that the pavement was from the Miocene period because it contained the remains of the three toed Miocene horse and thus makes the pavement thirty million years old. We might have to review all of our dating procedures as they are in apparent conflict too often than scientific and mathematical theory would allow.

I will now give you a question that you will see the answer to as you read on. Was there actually an Ice Age?

Did you know that the North Pole around fifty thousand years ago was centred at Hudson Bay in Canada and this appeared to be the centre of the North American glacial mass. Was it actually a glacial mass or the actual Polar ice mass? This is just something to think about here.

What we were taught in our schools and universities might not actually be true!

Around twenty thousand years ago the Yukon Teritory in Canada was supposed to be buried under a two mile thick ice mass. The Great North American ice mass. The jawbone of a domesticated dog was found at Old Crow in the Yukon Territory in Canada as well as the jawbone of an eleven year old child at a very old habitation site. Where was the two mile thick glacial mass that was supposed to have been here in this period? Oh, it was not here but it was elsewhere in Canada. So much for old science and assumptions? We will meet this ice free gap again as well as humans living in them. this was a very popular are for human habitation.

Around 19,700 BC cosmic dust obtained from ice cores at Camp Century in Greenland indicated levels of iridium and nickel with concentrations of one to two orders higher than those at present. These were obtained by neutron activation analysis. What do these mean? High levels of iridium generally indicate meteor bombardment. It is also amazing the level of error factor in these geological guesstimates as well. Once we allow for cosmic radiation fluctuations how correct are even these? What caused these increased iridium and nickel levels? Were they created by the 20,000 BC tektite storm? Hang on you say, how could you say that a tektite storm in the eighteenth millennium would effect iridium and nickel levels? Because there might be dating discrepancies. The impacts attributed to the eighteenth millennium might have occurred in the twentieth millennium or vice-versa. How correct are these ages again? We might as well just pick any date that we feel like!

At least now we have a meteor crater. Not just tektites or iridium anomalies. The Tenoumer Crater is 1.9 kilometres in diameter and was created around 19,400 BC, give or take nine thousand seven hundred years. The crater edges rise one hundred and ten metres above the base of the crater. Was this the impact crater that created the cosmic dust in Greenland? What about the tektites in Australia? The Tenoumer Impact Crater is in the Sahara Desert in Mauritania. There is that variable dating again.

Apart from the high iridium and nickel levels there was only one impact crater recorded which is the Tenoumer Crater in Mauritania recorded for the Twentieth Millennium BC. Remember that most of the surface of the earth is water and it has not been explored very much at all. Are there major impact craters in this period underneath the seas and oceans that are as yet undiscovered?

19 th Millenium BC.

There was nothing reported in the Nineteenth Millenium BC.

18 th Millenium BC.

Around 18,000 BC world weather conditions started warming up as glacial ice melted. World sea levels were one hundred metres or 328 feet lower than at present. World sea-levels then started rising abruptly. Some geologists stated that it only took five thousand years for the coastal plains to be flooded again as sea levels rose four hundred feet. What caused this enormous increase in sea levels and the massive decrease in glaciation? Sea levels rise and glaciation decreases. Sea levels lower and glaciation increases. Is this too simple an observation? This indicates that by thirteen thousand BC world sea levels were the same as now. Or were they? Geological guesstimates are pretty amazing.

Clube and Napier theorize that a giant comet settled into an earth-crossing orbit 50,000 years ago and for 30,000 years remained intact. 20,000 years ago a massive fragmentation event occurred and from 17,000 years ago multimegaton fragments may have periodically collided with the Earth thus causing a reduction in glaciation. There were also two large impacts in the eleventh Millennium BC and the ninth Millennium BC which raised global temperatures so much that the Ice Age was brought decisively to an end. Other researchers disagree with this and there are a lot of conflicting theories to explain the end of the last Ice Age. This was only one factor. I just add all of the research conclusions together and present them to you. The research conclusions can create their own combined results.

The astrophysicist Victor Clube at Oxford University proposed that a giant comet entered the Solar System some time before 18,000 BC and disintegrated over the next few thousand years leaving a debris strewn orbit in to which the Earth periodically blunders. The latest research has concluded that many asteroids, meteors and comets are actually pieces of rubble with or without ice pieces, such as in comets, and these pieces are held together by their own internal gravitational fields. That is until impact with something with a more powerful mass or gravitational field.

Victor Clube and John Napier state that after performing orbital analyses of several of the meteor showers that hit the Earth and after analyzing the paths of comets, asteroids and meteor showers they discovered that they were related to each other. Examples of these are the Taurids, Perseids, Piscids and Orionids as well as the comets Encke and Rudnicke and the asteroids Oljato, Hephaistos and at least one hundred others. They are all regarded as the offspring of one single massive comet that first entered our solar system 20,000 years ago or 18,000 BC. Clube and Napier calculate that due to subtle changes in the orbits of Earth and cosmic debris the Earth passes through the densest parts of the giant comet clouds every two thousand to four thousand years. Clube and Napier postulated that the Earth would pass through a four hundred year long cloud from 2000 AD on. I haven't seen it yet but anything is possible.

Clube has linked the cataclysmic meltdown of the last Ice Age between 12,000 BC and 7,000 BC with one such storm that caused sea levels to rise on average 100 metres or over three hundred feet. Imagine the impact on our coastal cities which generally are our largest and most prosperous ones. The same to a smaller degree occurred again around 3,200 BC, 2,350 BC and 500 AD.

My research indicates that this type of event also occurred several other times over the next few thousand years. In this period alone there were numerous meteoric impacts of massive size on the earth!

Microtektites called *australites* fell over a vast area of Australia twenty thousand years ago. There was also a massive tektite shower as well covering a very large strewn field at the same time. This strewn field is one tenth the size of the earth's surface. Other sources state that the australites also fell between seven hundred thousand and seven thousand years ago. These microtektites are linked to tektites found in Indonesia, Southeast Asia and the Philippines. Talk about geological dating variance! Geological dating variance is the discrepancies between different apparent geological dates for the same phenomena. It is almost arbitrary though the gelogists regard it as scientific and I will arbitrarily choose this date that some geologists have suggested for the *australite* fall. If geologists have given these dates then even if they are totally different they must be true.

And incidentally what are tektites and microtektites? Tektites are best explained as gravel-sized black, green, brown or grey natural glass formed from meteoric impacts or a meteorite storm. They are tektites when they are centimetres across and microtektites when they are millimetres across. What sort of aerial body or bodies shattered before impacting with the Earth into billions of pieces? Had the Earth wandered into a cosmos voyaging tektite swarm?

Geological studies indicate that the Great Barrier Reef off the coast of Queensland was an exposed landmass possibly only 20,000 years ago around 18,000 BC.

There are rock drawings of Gods apparently in bulky suits and helmets on the walls of the Yarbiri Soak in the Northern Territory. They must be over twenty thousand years old as they are spread over rocky clefts that have been broken up and hollowed out by erosion during this period. Even if they are not gods they are still incredibly old. The Yarbiri Soak is near Yuendumu in the Northern Territory in Australia.

The Veevers Impact Crater is twenty metres wide rising one and a half metres above the surrounding desert plain. The deepest point of the depression is seven metres below ground level. The rim to rim diameter is around seventy metres. Iron meteorite fragments have been found here around the crater. The impact is twenty thousand years old but other sources say six thousand years old. Always conflicting dates. The Veevers Crater is in the Pilbara region in Western Australia in Australia.

Around 20,000 years ago the average sea levels were three hundred to four hundred feet lower than they are today and areas like Beringia which was between Alaska and Siberia are now above sea level. Beringia is now the Bering Strait. The Russians hypothesize that Beringia was a steppe on the Siberian side that became tundra on the American side. The grassy flora of Siberia resembled that of East Africa and only started to resemble tundra on the American side. Of the thirty four species of extinct animal in Siberia twenty eight of them were from temperate climatic zones. Alaska was largely icefree whilst Siberia was completely ice free. Was Siberia more like a savannah rather than a steppe?

The Iturralde Crater, also called the Araona Crater, in the Amazon rainforest area of Bolivia was created twenty thousand years ago, others say eleven thousand years ago, even others twelve thousand years ago, and is an eight kilometre in diameter circular impact crater. Do I need to say it again?

Iiri Chiachula, a researcher in Quaternary Geology at the University of Alberta, suggests that there was very early human activity under what was later glacial Lake Calgary in Alberta in Canada. Stone tools were found under glacial till dating from 18,000 BC. The tools were twenty metres down. The tools included flakes, cores and pebble tools. Where is the two mile thick ice mass again? Like in the Yukon Territory previously the ice mass was still missing.

Lake Lasnamae in the city of Tallinn in Estonia is an impact crater that is twenty thousand years old and half a kilometre in diameter. Do you know how many major impacts leaving craters at least half a kilometre in diameter have there been in our last millennium? That you know of? You know of Tunguska. That is one. Are there any more? Now you are stretching it. Keep reading. This was an unusual millennium and more would come again.

At San Michel d'Arudy in Aquitaine in the Basses-Pyrenees of France a most unusual carving was found. There is a carving on bone here of a horse's head wearing a bridle over seventeen thousand years before the supposed invention of the bridle and several thousand years before the accepted domestication of the horse. This is our fourth French bridle and from the same period. One of the others was found at La Marche at Lussac.

Around 18,000 BC world weather conditions started warming up as the ice melted. Warmer and wetter weather conditions in Greece allowed trees to grow over the landscape again. The landscape was previously sparsely vegetated. Open oak forests as well as beech, holly, hornbeam, pistachio and almonds took over at low altitudes as fir and pine forests took over the higher regions. This indicated that glaciation was retreating. Watch the trees and they will show you the climate.

What are these climate reports to do the main theme of this book? They are indicators of massive change of climate and of the Ice Age itself and its variations.

The remains of fossils of camelids, giant armadillos and extinct horses have been found in the cenotes or underground caves of Yucatan in Mexico. This would indicate that in this period the area was covered with dry grasslands. Now it is jungle.

Skulls have been found in the Andes of Peru of an unidentified race whose remains are at least twenty thousand years old from 18,000 BC and are found in tombs with mastodon bones. Mastodons looked very similar to mammoths and inhabited the Americas up until the end of the last Ice Age whenever that actually was. Mastodons had shorter legs, a longer body and were more muscled than mammoths being very similar to the modern Asian elephant. The mammoths were also similar to the Asian elephant as well. All of them had prehensile lips to eat leaves off trees. African elephants are grass eaters generally with a non prehensile or extendable lip structure.

In the Sierra Morena in Andalusia in southern Spain there are alphabetical looking symbols in the local caves that predate the accepted invention of the alphabet by many thousands of years. These symbols are 20,000 years old and undeciphered as yet.

There is an impact crater in Lake Tremorgio in Switzerland. The crater is 1.36 kilometres wide and was formed around twenty thousand years ago. Other sources state fifty thousand years ago. We had a report of another impact event in Switzerland only four thousand years before at Lake Constance. An amazing coincidence? The two lakes are only eighty-five miles apart.

Around 20,000 years ago as the sealevels started rapidly rising at the same time a continental expanse began to sink beneath the waves in Southeast Asia. This was the Sunda Shelf or Sundaland and it lay where the southern reach of the South China Sea, the Gulf of Thailand and the Java Sea are all now. A land area equivalent to the Indian subcontinent vanished below the sea leaving only the Malay Archipelago, Indonesia, Indochina and Borneo protruding from it. A large section of land along the Pacific coast of East and Southeast Asia was submerged as well. At present these seas are only several hundred feet deep. Originally this area was probably inhabited but nothing is known of who or where except the dimmest of legends. As you read on you will read other dates for the sinking of Sundaland. It might have developed in stages or suddenly. As you read on you might find that it actually was later but I am still going to introduce each item to you when geolgists insist, though you are going to find these same geologists disagreeing as well.

The remains of a man made wall have been discovered off the west coast of Taiwan between the small islands of Don-Jyu and Shi-Hyi-Yu by Professor We Miin Tan. There are legends of a submerged town here called Mudala but in 2002 only a wall had been discovered. The wall is one hundred metres long and is 28 meters below sea level. The wall is fifty centimetres wide, one metre tall and is positioned on an east west axis. The wall is believed to be twenty thousand years old. The two islets are near Penghu in Taiwan. Believe it or not but were there other cultures on Earth before the Cataclysm at the end of the last Ice Age?

In 1969 a group of scientists under Professor Leonidov Marmajaijan discovered a cemetery in a cave in Ashkhabad in what was then Soviet Central Asia or Turkmenistan. In one large grave were thirty skeletons in a perfect state of preservation. They were radiocarbon dated as 20,000 years old from 18,000 BC. Eight of the skeletons showed serious bone injuries made when the subjects were still alive, including claw and bite marks. One skeleton showed traces of a surgical operation on the chest area where the ribs on the left side of the skeleton had been cut away, an opening had been made and widened by retraction to permit an operation. After the success of this operation the patient survived another five years as is shown by the thickness of the periosteum and subsequent bone regrowth. The rib cut was identical to the cardiac window used today. Some authorities state that these operations actually happened 100,000 years ago. I am not sure about one hundred thousand years ago though but I am not an expert in dating these things. I only bring you the reports. You do what you want with them.

Topper is in Allendale County in South Carolina. Topper appears to be an industrial scale toolmaking site that was in existence 16,000 to 20,000 years ago. Other sources state that evidence is dated to 50,000 years ago. Here are those differing dates again. Abundant Clovis-style chert artifacts have been found here.

The Elko Crater in Elko County in Nevada in the United States is twenty thousand years old and 250 metres in diameter.

How many meteorite impacts did we have in the Eighteenth Millenium? We have Veevers Impact Crater in Australia, the Iturralde Crater in Bolivia, Lake Lasnamae in Estonia, Lake Tremorgio in Switzerland and Elco Crater in Nevada. Is this unusual? And these are only the found ones. This is five large impacts in one thousand years. Is this still quite a lot? And don't forget the Australites and tektites showering upon Australasia as well as Southeast Asia, Indonesia and the Philippines.

17th Millenium BC.

The Meadowcroft Rock Shelter is on Cross Creek, a small tributary of the Ohio River, in Washington County in Pennsylvania. Pre-Clovis period artifacts have been found dating to the period of 11,950 BC to 12,550 BC. Meadowcroft is near Avela in Washington County. Some radio-carbon dates suggest that Meadowcroft dates back to 17,000 BC. Items that have been found here in what some describe as the oldest Paleoindian site in North America are ceramic items, bifaces, bifacial fragments, lamellar blades, a lanceolate projectile point and chipping debris. A hearth indicating the use of fire has also been found from this period. Another report states that in the Meadowcroft rockshelter near Pittsburgh James Adovasio and his team in 1973 discovered eleven distinct occupation layers dating back to 10,800 BC though burnt wood from one floor level gave a date of 13,000 BC and one very deep level containing the burnt remains of matting or basketwork dated back nineteen thousand years to 17,000 BC. This was a very popular area. Love those varying dates.

Clovis Points are from ten to twelve centimetres long, 4.8 inches, and date back at least to 8,000 BC to 11,000 BC. They were able to penetrate the skull bone of a mammoth. First found in 1932 these thin flint blades were first discovered in Blackwater Draw near Clovis in New Mexico by a highway construction company. These were fluted flint spear points that were larger and longer and in fact earlier than the Folsom spearpoints.

The tally of large meteorites to strike the earth in the Seventeenth Millenium is nil. Why so different to the Eighteenth Millenium? Why do some Millenia have nil impacts over half a kilometer in size and others have a plethora?

16th Millenium BC.

World sea levels are now 129 metres or 425 feet below present levels. This was a drop of 29 metres or 97 feet over only two thousand years. What is with this constant up and down of sea levels? How erratic was ice creation? Sea levels go down when ice masses grow and sea levels rise when the ice masses melt. Another report states that around 16,000 BC world sea levels were 142 metres or 466 feet below present sea levels. This indicates massive glacial ice build-up at the time.

Take your pick with these two as they are both from reliable scientific sources.

The period of 16,000 BC was the fourth in a cyclical rhythm of Ice Ages that had occurred since approximately 355,000 years ago. The next was 245,000 years ago followed by 135,000 years ago and finally 18,000 years ago. There are two periodicities involved, 100,000 years and 41,000 years. Accepted theory is that this could indicate that the changes to the orbital parameters of the Earth cause Ice Ages and the eccentricity, obliquity and precession of the axis cause variations in the intensity and distribution of solar radiation. This is documented by Vostok ice cores as well as Greenland ice cores.

The Vostok and Greenland ice cores indicate that there were massive changes in atmospheric concentrations of CO_2 (carbon dioxide) and CH_4 (methane) which are the most important greenhouse gases. All four climatic transitions indicated by Vostok ice cores show that the transition from glacial to warmer periods were accompanied by sudden increases of 180 to 300 parts per million of carbon dioxide and equally sudden increases of methane from 320 parts per billion to 650 parts per billion. The Greenland ice cores show that shifts in methane levels coincide with rapid and major changes of temperature in the Northern Hemisphere. The huge glaciers of this period had absorbed so much water and had placed so much weight onto the Earth's crust that sea-levels on average were more than ninety metres lower than modern levels. Maximum glaciations in North America reached as far south as thirty-nine degrees north. What caused the ice masses to change so suddenly?

The Aegean Sea rose three hundred feet between 17,000 BC and 16,000 BC. When in this thousand year period did the sea rise from 425 feet below present levels to 125 feet below present levels? Was this another enormous and very fast glacial melt?

Russian Scientists concluded that the Arctic Ocean was warm during most of the last Ice Age and the period 32,000 to 18,000 years ago as being particularly warm. Here are those vanishing ice masses again. Sea levels indicate increased glaciation. Which one is right?

There were two enormous ice sheets that mantled all of Canada and the northern portion of the United States. One of these was the Laurentide Ice Sheet that covered eastern and central Canada with its centre in northern Quebec, Labrador and Newfoundland and spreading southwards to Pennsylvania, Ohio, Indiana and Illinois. The other ice sheet was the Cordilleran that was on the west coast and started in British Columbia down to the latitude of Seattle in Washington State in the United States. The Cordilleran covered much of the Pacific Coast though not all of it leaving large pockets of ice-free land.

In this period the North Pole was in the position where Hudson Bay is now. This was not unnatural glaciation but normal ice expansion from a Polar region. The Albedo Effect would determine how large it would grow and cometary bombardment would determine where the Poles would then be. The Albedo Effect is where ice and snow reflect back more of the sun's energy and absorb less. The air temperature decreases and ice and snowfields grow until competition with negative feedback mechanisms causes equilibrium. The reduction in forests also increases albedo. This is theory. What if the feedback mechanism is actually a cometary or meteoritic bombardment that knocks the Earth off its axis and into a new one?

What are glacial periods? Glacial periods are periods of cooler and drier climates over most of the Earth with large land and sea ice masses extending outwards from the Poles. The snow line is lower so mountain glaciers occur in unglaciated areas and at lower altitudes. Sea levels also drop due to accumulation of ice in the ice caps. Then ocean circulation patterns are changed. Whilst this is occurring positive feedback processes occur wherein the Albedo of the Earth is influenced. Hear me out on this as you read on.

In the Sixteenth Millenium glacial ice sheets spilled into England from Scotland.

There was no English Channel and *Cro-Magnon* man could easily walk from England to France.

Yes, they are all disparate items of information. But they start to assemble a history which you will find unbelievable. You will also find it more unbelievable that you were never taught about this hidden history. You were betrayed by modern paradigms and education systems preserving their own theories at the expense of evidence.

Much of Northern Europe was uninhabitable Polar Desert with cold, dry shrub tundra skirting it. The glaciers were advancing south though. People were forced into southern France or the Central Russian Plateau. No humans lived between the Scandinavian Glaciers and the Alpine Glaciers as there was no ice-free corridor between them.

The climate of southern France was now reasonably hospitable. Pollen studies show that trees now grew in sheltered valleys and the fauna included mammoth, woolly rhinoceros, horse, bison, aurochs and many types of antelope. This was regarded as a refuge area from glaciation. Evidence from the Vezere River Valley in the Dordogne in France in this period the winters lasted an average of nine months in France.

At the same time glacial ice sheets spilled into Germany from the north.

Meanwhile in Asia some of the earliest pottery in the world was discovered in Ishigoya Cave near Nagano in Honshu in Japan. Pottery jars were in use here around sixteen thousand BC. Pottery for making vessels and pots etc was not supposed to have been invented yet. Obviously no one had told the Japanese who are still masters of pottery making in the Twenty-first Century.

There was a massive meltdown of the ice sheets in the Northern Hemisphere and the ice sheets rapidly retreated. Was this at the beginning or the end of the Sixteenth Millenium?

There was a massive meltdown of the ice sheets in the Northern Hemisphere and the ice sheets rapidly retreated. The area now occupied by what is now the North Sea was a steppe area and the Baltic Sea did not exist at all.

When during the Sixteenth Millenium did this occur? It could not have been when the sea levels were still lowering. There must have been a time when the sealevels started rising again.

The river valleys of the Central Russian Plain were rich in trees in sheltered valleys and numerous animals ideal for hunting abounded such as mammoth, woolly rhinoceros, antelope, horse, bison and aurochs. Where are the glaciers? The tundra was retreating.

Vast ice sheets up to four kilometers thick still covered Scandinavia and Scotland. All of Scotland was covered with glacial ice up to four kilometers thick.

At Cactus Hill on the Nottoway River near Blackstone in Nottoway County in Virgina Joseph M. McAvoy and his wife Lynn McAvoy discovered evidence of human occupation dating back to 16,000 BC. This is much earlier than Clovis. Clovis culture tools had been found above the settlement layers of this older culture. This earlier culture hunted white tailed deer and mud turtles.

The tally of large meteorites to strike the earth in the Sixteenth Millenium that we know of was nil.

15th Millenium BC.

The present ice that covers Antarctica, the South Pole, appears to have been only formed around the period of 4,000 BC. Before that, as far back as 15,000 BC, the end of the last Ice Age, the area was ice-free.

The North American Ice Cap disappeared around the same time when the Arctic, North Pole, began to freeze over. Charles H. Hapgood suggested that the Earth's crust itself slipped, pushing North America closer to the Equator and Antarctica further away towards and over the South Pole. Hapgood believed that the weight of ice in the Polar Regions caused the slippage. What happened though when the ice melted causing the three hundred foot high rise in sea levels that occurred at the end of the last Ice Age?

Could bombarding comets or meteorites have caused the sudden tilting?

Could the crashing comets and meteorites actually cause the glacial melting?

Some of the islands of the Arctic Ocean were never covered by ice in the last Ice Age. On Baffin Island which is nine hundred miles from the North Pole alder and birch tree remains found in peat suggest a much warmer climate less than thirty thousand years ago. These conditions prevailed until seventeen thousand years ago or 15,000 BC.

During the Wisconsin Glaciation or Wisconsin Ice Age there was a temperate climatic refuge in the middle of the Arctic Ocean for the flora and fauna that could not exist in Eastern Canada and the Northern United States.

A Soviet expedition removed rock from the north of the Azores from a depth of two thousand metres down and found that these were formed above water 17,000 years ago in 15,000 BC. Were these from former land masses in the Atlantic? Were the land masses lifted above sea level or were sea levels rising on their own.

Fairbanks studied coral reefs off Barbados that had been growing there for twenty thousand years. Fairbanks noticed that one particular type of coral grows only in shallow water and dies if the water gets too deep. Cores through the coral layers showed sudden changes in sea levels in this period indicating that the ice masses had melted quite rapidly and not gradually as originally theorized.

There were two major jumps in radiocarbon levels around 15,000 BC and then around 11,000 BC. Fairbanks insists that this was a sealevel rise of fifty feet in a few weeks. This is the same as cores taken from Tahiti and the Sunda Strait that also indicate 11,000 BC. And don't forget to allow for geological dating variation.

Around this time the Black Sea was a large brackish inland sea that was called by geologists the Euxene Lake. The Euxene Lake was not connected to the Mediterranean Sea which might not have even existed then at all either. In this period world sea levels were down to three hundred feet below current sea levels. Many researchers state that in this period the Mediterranean Sea was a series of massive lakes in a huge valley. All of our continental shelves were above sea level and all would be completely flooded at the end of the last Ice Age.

What is with the archeological reports? They indicate that there were types of civilization when we thought that there were only Neanderthals wandering around. In 15,000 BC major engineering feats were being accomplished at Glastonbury in Somerset in England. Aerial archaeology has discovered a vast zodiac, visible only from the air with Glastonbury Abbey, originally an island, at its centre. The zodiac is ten miles across and if one theoretically raises the sea level in this area to what it was in precataclysmic times then the ten figures of the original zodiac are clearly visible from the air. Glastonbury Tor, an incredibly ancient man-made hillock with a spiral pathway leading to the summit is at the centre. Why is the entire complex only visible from the air? Why are so many ancient structures only visible from the air?

All over the world strange structures have been found from this period. In the Gorge of Chabbe or Sappe in Tember in Ethiopia which was originally a cavern there is a processional carving of fifty headless cattle with horns coming out of their necks and all pointing towards where the entrance of the cavern would have been. Around 15,000 BC the cavern roof collapsed partially burying the works and this only gives us the latest possible date for the execution of the works as they could even be older.

You never know what you will find or how it will be interpreted. In the caves of Les Trois Freres near Montesquieu-Avantes in the Midi-Pyrenees of France there are cave drawings of funny little flying cars that look exactly the same as the flying cars in the television series the Jetsons. There are even tiny little television aerials sticking out of the tops of some of them. Where in nature in the Magdalenian period, 15,000 BC, would these types of things be seen? Remember these are the same artists that created the incredibly realistic cave paintings depicting cattle and other realistically depicted animals. What? They were really good with animals and abbysmal with rigid objects?

In the cave system of Niaux there is a depiction of two flying discs that look like cartoons, one of them even trailing a wavy dotted line behind it to indicate movement. Niaux dates from the Magdalenian period of 15,000 BC. Niaux is in Ariege in the Midi-Pyrenees of France. Both sites are only twenty-four miles apart. Were these images part of a local culture or not? Were they flying craft or meteoric phenomena?

Soviet Geologists prospecting in the Tien Shan Mountains in Kyrgyzstan in almost inaccessible mountains were astonished to find heaps of slag and well-worn picks, galleries and pit shafts dating from the Upper Paleolithic Age, around 15,000 BC. This shows very technical iron ore mining long before it was supposed to be around. To add insult to injury the inhabitants used bronze picks well before the beginning of the Bronze Age as well. Bronze is a combination of copper and tin. Yes, there is a bibliography at the end of this book. There always is in my books.

In the La Cullalvera Cave in Ramales de la Victoria in Cantabria in Spain there are cave paintings that are tens of thousands of years old that depict flying craft that could resemble flying saucers or aerial constructions. These cave drawings could depict flying saucers of all shapes and sizes and with the utmost realism. These drawings are of the Magdalenian period of 15,000 BC. More Magdalenian UFOs? Or ritual objects as they are often called?

In the caves of La Pasiega there are drawings of what appear to be flying saucers dating from the Magdalenian period of about 15,000 BC. There actually is nothing else in nature that they would look like even with a very vivid imagination. La Pasiega is also in Cantabria near Puente Viesgo in Spain. La Pasiega is only eighteem miles from La Cullalvera. The two sites in Spain are only 380 kilometres from the two sites, Les Trois Freres and Niaux, in France. Were they ritualistic drawings from one culture?

Cores through coral layers in Tahiti taken by Bard and colleagues showed sudden changes in sea levels in this period indicating that the ice masses had melted quite rapidly and not gradually as originally theorized. There were two major jumps in radiocarbon levels around 15,000 BC and 11,000 BC. Fairbanks insists that this was a sealevel rise of fifty feet in a few weeks. This is the same as cores taken from Barbados and the Sunda Shelf in the China Sea that also indicate 11,000 BC. How would modern civilizations handle these sudden raisings and lowerings of sea levels? Fifty feet in only a few weeks? We whinge and scream about several millimetres a year in the current age. What caused the sudden sea level changes? Were they melting ice masses or tsunamis from earthquakes or massive meteoric impacts into the oceans or as you will see possibly all of the above.

Human remains were found at Laguna Beach in Orange County in California. A human skull was found here that was seventeen thousand years old. The skull was called the Laguna Beach Woman and for a time was regarded as the oldest human fossil remains in North America after its discovery in 1933. We will just try and ignore the older ones will we?

Around 15,000 BC a glacial ice dam that had blocked the Clark Fork River in Montana created Lake Missoula. This dam had failed at least forty times and caused horrendous floods. One of the Lake Missoula floods over where Portland in Oregon is now would have covered it in four hundred feet of water. What was causing this sudden melting of glaciers and ice masses? Gradual melting does not cause sudden massive land tsunamis!

Ice Age dams filled with glacial meltwater repeatedly failed and flooded and eroded land in Washington State and Oregon. Some of these onrushes of water exposed petrified logs in Gingko Petrified Forest State Park near Vantage in Washington State.

The remains of a pre-Clovis mastodon kill site dating to 14,510 BC have been found at Saltville in Smyth County in Virginia. The prehistoric inhabitants also utilized parts of a musk oxe and left biface traces. Cactus Hill is only three hundred and thirty kilometres from Saltville.

In the Paisley Caves in Lake County in south-central Oregon coprolites or human faeces were found as well as tools, thread, cord and baskets. The fossils were found in the Paisley Five Mile Point Cave. A small rock-lined hearth has been found here as well as the bones of waterfowl, fish and large mammals including camels and horses. Coprolites for those who do not know are fossilized faeces. These date back to 14,510 BC.

Cosmic dust from ice cores taken from Camp Century in Greenland indicated levels of iridium and nickel with concentrations of one to two orders higher than those at present. These were obtained by neutron activation analysis. High iridium levels indicate meteoric bombardment in this period. It is also amazing the level of error factor in this geological guesstimate as well. Once we allow for cosmic radiation fluctuations how correct are even these?

There were no recorded major impacts during the Fifteenth Millenium BC so why the iridium spike? Are our dates all a little confused?

The tally of large meteorites to strike the earth in the Fifteenth Millenium is nil. Maybe they all fell into the sea?

That could explain sudden and dramatic sea level rises as indications of how long these sea level rises lasted is unknown. They could well have been only temporary.

14th Millenium BC.

This was the advent of the Bolling Interglacial phase of warm climate in Europe and the Brady Interstatial in North America.

The Bolling Interglacial started around 14,000 BC and lasted until 10,300 BC when the Ice Age suddenly returned in the age called the Younger Dryas wherein Ice Age conditions were nearly as before the Bolling Phase.

From around 80,000 years ago an immense ice cap with huge glaciers reached deep into Europe, Russia, Canada and the United States. This ice cap was more than a mile thick in the north, covered all of Ireland, most of England as far south as London, and stretched across Europe. In North America an ice cap almost two miles thick reached as far south as Saint Louis and Philadelphia and further south still were endless plains of Arctic tundra.

The glacial sheet in the Southern Hemisphere moved from the present tropical regions of Africa towards the South Polar Region.

In the Northern Hemisphere the ice in India moved from the Equator towards the Himalayan Mountains and the higher latitudes.

The glaciers of the Ice Age covered the greater part of North America and Europe while the north of Asia remained free, interestingly enough the home of the woolly mammoths. In America the plateau of ice stretched from latitude 40 degrees and even passed this line. In Europe it reached latitude 50 degrees. In northeastern Siberia, above the Polar Circle, above latitude 75 degrees, there was no perennial ice.

Tilt all of this sideways and you have the North Pole in North America! Exactly where Hudson Bay is now.

Most of Europe had been buried under ice two miles thick. So too was most of North America where the ice cap had spread from near Hudson Bay to enshroud all of eastern Canada, New England and much of the Mid-West down to the 37th parallel, well to the south of Cincinnati in the Mississippi Valley and more than half way to the Equator.

When the North Pole was at Hudson Bay the ice cap strangely enough extended south as far as Ohio and did not cover some of the islands of the Canadian Arctic Archipelago north of Hudson Bay. Also the Yukon District in western Canada and northern Greenland were not ice covered either. The ice sheet was thicker and extended further south on the low central plains of the Mississippi Valley where it covered Wisconsin and Ohio than on high mountain areas to the west though at the same latitude. This is the opposite of normal glacial theory which states that glaciers form on mountains and then head down. These last Ice Age glaciers did not even go near the western mountains.

Allowing for the North Pole being in the position of Hudson Bay then the northernmost Arctic islands would then be south of the North Pole by one thousand miles and a similar effect was in regard to the western mountains.

At the Ice Age's peak seventeen thousand years ago the total ice volume covering the Northern Hemisphere was in the region of six million cubic miles and there were extensive glaciations in the Southern Hemisphere as well. The world's seas and oceans had provided the surplus water flow from which these numerous ice caps were formed, which were then about four hundred feet lower than they are today.

Geologists refer to the sudden and miraculous sudden meltdown as the Bolling Phase of warm climate in Europe and the Brady Interstatial in North America. In both regions an icecap that took forty thousand years to develop disappeared in only two thousand years. Possibly even less if you allow for dating inaccuracy. This could not have been the result of gradually acting climatic factors that normally explain the end of Ice Ages. The rapidity of the deglaciation suggests that some extraordinary factor was affecting the climate. The dates suggest that this factor first made itself felt about 16,500 years ago around 14,500 BC and that it had destroyed most, perhaps three quarters of the glaciers by two thousand years later around 12,500 BC, and that the vast bulk of these developments took place in a millennium or less.

What happened in the Chaleux Cave in Namur Province in Belgium? The cave of Chaleux was buried by a mass of rubbish caused by the falling in of the roof in the remote past. The split bones of mammals, mainly horses and to a lesser extent reindeer, and the bones of birds and fishes were found here as well as an immense number of objects, chiefly manufactured from reindeer horn, such as needles, arrowheads, daggers and hooks. Beside these were ornaments made of shells with engraved figures, pieces of slate also with engraved figures, mathematical lines, remains of very coarse pottery, hearthstones, ashes, charcoal and thirty thousand worked flints mingled with the broken bones. In the hearth placed in the centre of the cave was discovered a stone with certain but unintelligible signs engraved on it. These remains are from well before the accepted invention of writing many thousands of years later. When was writing invented? The Chaleux Cave remains date from the Bolling Interglacial.

A sixteen thousand year old drawing of a horse in the caves at Lascaux has been found to be a carefully recorded lunar calendar. It was discovered by Dr Michael Rappenglueck of the University of Munich. This was in the Lascaux Caves near Montignac in the Dordogne in France. How primitive were early people?

Rock art found in the Coso Mountains in Inyo County in California dates back sixteen thousand years.

There is evidence that ice-free plains extended along the British Columbia Coast of Canada which are now at a depth of 130 metres below sea level around 13,700 BC. There was a diverse ice-free eco-system here that could have been conducive to migration.

Manwhile on the other side of the Atlantic Ocean more was unfolding. The Pedra Pintata is near Tarame in Roraima State in Brazil on a plateau that extends from the Rio Urari Coera to the Sierra Pacaraima. The stone stands in the middle of an immense plain and is visible for miles. It resembles an egg of enormous size. Pedra Pintata means Painted Rock. In 1940 Professor Marcel Homet discovered a gigantic stone egg 328 feet long and 98 feet high on the upper Rio Branco in North Amazonas. There are countless crosses, letters and Sun symbols over a surface area of seven hundred square yards. Homet believed that the rock was a manmade work carried out by stonemasons over many decades. The middle course of the Rio Branco is the Parima. Other sources state that the stone is 390 feet long, 250 feet wide and 90 feet in height. The egg shaped stone is covered in hieroglyphics, and ancient carvings and rock paintings including the picture of an enormous serpent on the rock's side, turtles and symbols of Sun worshippers. Chemical analysis has proven that these are thousands of years old. Homet found engravings that indicated that the ancients knew arithmetic. Instead of a decimal system there were engravings depicting 3, 5, 7, 9 and 12. Homet even discovered sketches of horses, wagons and wheeled vehicles. The aged patina around the paintings indicated that these sketches were thousands of years old. There is a platform under the stone. The dolmen contained five artificially hewn angles engraved with signs of an ancient megalithic race. The shape of the dolmen and the symbols were similar to those he had found in South Africa. Homet discovered other dolmens as well around the rock. Behind one there were four tomb grottoes cut into the rock. A tunnel led up through the centre of the rock towards the top. The serpent on the side measures over twenty-two feet long and commands thousands of inscriptions and letters. There are images on the rock of horses, wagons and wheels many times repeated. They are always drawn in profile. There is an unexplored passage in the rock that travels to a large chamber in the tip that according to the local Indians was where victims were suffocated by poisonous gases from the bowels of the Earth. One side of the great rock had been hollowed out and there were many passages filled with rubble. The caves were full of human bones. There are stratified Paleoindian deposits here with radiocarbon dates for burnt plant samples here being given of 11,000 years. This is contemporary to the Clovis culture of North America. Triangular spearpoints found here though are not Clovis-like and indicate a different culture. Thermoluminescence dating indicates a date of 16,000 years old or 14,000 BC.

    The site of Taimataima is in the Caribbean coastal region of Venezuela just east of the Isthmus of Paraguana. Remains have been found here of an indigenous culture that had developed tools and weapons earlier than Clovis in New Mexico. This was around 13,500 BC. If the first arrivals in America had come from the north via Beringia and had reached Clovis around 11,500 BC, who were the Taimataima people and where had they come from? Taimataima was a mastodon kill and butchering site in Venezuela. Hang on? This was not a snowy area! What were mastodons doing here? Don't tell me that they were temperate climate animals? In fact so was Siberia in this period! The lithic projectiles found here were technologically different to the North American Clovis fluted point projectiles. Who are all these different people inhabiting the Americas well before the accepted dates of migration? As well if they had come from the north why are they concentrated in the south?

    By 13,528 BC at Buttermilk Creek near Salada in Bell County in Texas are more precataclysmic relics. There is evidence here of a culture predating Clovis. Flintknapped chert nodules were worked into bifaced preforms that became spear points, knives and other tools. Who were these people all over North America where it was unglaciated?

    At Cactus Hill on the Nottoway River in Nottoway County in Virgina Joseph M. McAvoy and his wife Lynn McAvoy discovered evidence of human occupation dating back to 16,000 BC. This is much earlier than Clovis. Clovis culture tools had been found above the settlement layers of the older culture. This earlier culture hunted white tailed deer and mud turtles. Other scientists state that this was around 13,070 BC. This is why we have both dates represented here. Never let it be said that I am ignoring traditional dating techniques and discoveries. Some researchers have theorized that the stone tools found here were a cross between Clovis and European Solutrean. This group of colonists did not come via Alaska either? Did they come from Europe via island chains across the Atlantic Ocean whilst sea levels were much lower? Were they actually European Solutrean colonizers?

The meteorite score for the Fourteenth Millenium is nil for large impacts more than half a kilometre in diameter. The winner so far is the Eighteenth Millenium.

13th Millenium BC.

The Kreuz comets are long period comets that come in close to the Sun, being visible only during eclipses. This family of comets was originally one large comet that broke up approximately 15,000 years ago, or in 13,000 BC. Is there a connection between the Kreuz comets breaking up and the end of the last Ice Age? This is the middle of the Bolling Interglacial Period.

Is there a connection between the Kreuz Comets and Tollmans comet as well as Clube and Napiers comets?

In the period of 13,000 BC there was a massive peak in radiocarbon in the atmosphere of our planet. This could be caused by a supernova or explosion of a massive star when there is a pulse of radiation hitting the Earth's atmosphere that creates a surge of radiocarbon. This can also be caused by a solar flare. A third probable cause could be a declining of the Earth's magnetic field or a thinning of the atmosphere. Another cause can be comets or asteroids impacting with the Earth. Is this tied in with the Kreuz Comets?

May I state that these data facts are all from different as well as independent published scientific sources.

About 13,000 BC there was a low sea level of 425 feet below the present level. From this date there was a rapid rise until about 5,000 BC. The sea level changes were accompanied by tectonic violence in the Earth's crust including volcanic eruptions. Other sources state that this sea level rise was much faster.

Suddenly in 13,000 BC climate warming accelerated dramatically. This caused massive change to hunting patterns in Europe as Ice Age animals such as mammoth and bison, arctic fox and reindeer suddenly migrated northwards as the tundra rapidly retreated to be replaced by birch and forests of deciduous trees. The climate warmed rapidly, there were higher summer temperatures, the seasons were more clearly marked and the winters were less severe. Research on the lips of frozen Siberian rhinoceroses and mammoths indicate that they had long prehensile lips which are designed for eating trees and shrubs the same as Indian elephants. African elephants have short lips for eating grasses generally only supplementing this with leaves from trees and shrubs occasionally. So much for the frozen Siberian tundra dwelling mammoths. They would have starved to death.

Russian researchers estimated that prior to their extinction around 9,000 BC in Siberia there were over eleven million of the great beasts covering an area of 575,000 square miles. There were also an equal number of woolly rhinoceroses and bison and horses. These creatures cannot live on snow and moss. These animals required several hundred pounds of food per day. This does not exist in perma snow nor does freshwater which they would also require. They would have died out long before this. These animals are buried along with the mangled remains of destroyed forests that appear as if massive tidal waves had buried them. These were the destroyed forests that they and their other animal fellow residents resided in, ate from and became trapped in when the mud tsunamis started.

After 13,000 BC there was a rapid increase in acorn-rich oak forests in eastern Iran, the Jordan Valley and other locations in the Middle East. Surface water was abundant and freshwater springs provided ample drinking supplies throughout the area. The climate was warmer and far more moist than it is now.

From 13,000 BC to 11,000 BC there was a major increase in rainfall in the world. This was followed in 11,000 BC by a tremendous drought that was possibly triggered by the sudden overflow of Lake Aggasiz in North America.

Around fifteen thousand years ago, around five thousand years after the coldest period of the last Ice Age, the earth's climate had been rapidly warming. This was when what is called the Heinrich 1 Event occurred. Heinrich Events are when there is a massive discharge of icebergs from North America into the Atlantic Ocean and instead of just plankton being the main composite of them instead there is glacial debris such as stones and dust that has been carried far out to sea. Heinrich had originally discovered six layers of tiny white stones from North America in sediment cores taken from North Atlantic seamounts. On at least six occasions during the last sixty thousand years the relative portion of fine pebbly debris spiked dramatically. Later research by others indicated that these Heinrich Events were in fact widespread in the North Atlantic. The Heinrich layers were thickest to the north and west towards Hudson Bay in northern Canada and each was deposited very rapidly when the ocean was exceptionally cold. The ice in Hudson Bay had built up over several cold and warm oscillations, called Dansgaard-oeschger Oscillations, and the oscillations grew progressively colder as the cold-based ice-sheet in Hudson Bay grew. The theory is that eventually the ice became thick enough to trap some of the earth's heat which thawed the base resulting in stones, mud and water created by the thaw lubricating the ground underneath so that the ice skated across the underlying bedrock dumping icebergs and the debris into the North Atlantic. In this way Hudson Bay removed accumulated ice. Incidentally the Hudson Bay ice sheet acted in this way yet the Laurentide ice sheet did not. Was there some other causal factor as well? What happened next is that the sudden introduction of millions of gallons of glacial freshwater into the Northern Atlantic Ocean shut down the circulation of warmer water in the Gulf Stream which depends on the downwelling of salt water in the Labrador Sea for its existence. This would result in a deep freeze in Europe as the prevailing warm westerly winds would cease. The world would suddenly get dryer because the sudden cooling would reduce the amount of water vapour in the air as storm tracks moved southwards. This was the last Heinrich Event.

When the last Heinrich Event occurred there was an abrupt retreat of the Laurentide Ice Sheet as rapid warming occurred. Both the Cordilleran and Laurentide ice sheets rapidly retreated after Heinrich 1. The Laurentide retreated north and east into Sub-Arctic Canada and the Cordilleran shrank with amazing speed into the mountains of the western portion of North America. The only vestiges of the Laurentide Ice Shelf are the Great Lakes that were formed four thousand years after the retreat. A massive ice free corridor opened up allowing human migration from the northwest to join up with the people already resident there apparently.

In North America Lake Agassiz (Manitoba and North Dakota), Lake Chicago (Lake Michigan) and Lake Maumee (Lake Erie) all drained into the Mississippi River. Lake Agassiz, Lake Chicago and Lake Maumee were all massive glacial lakes.

Concerning Ross Sea sediments in Antarctica Dr Jack Hough of the University of Illinois showed that the log of core N-5 shows glacial marine sediment from the present to one hundred and seventy thousand years ago. From six thousand to fifteen thousand years ago, the sediment is fine grained with the exception of one granule at about twelve thousand years ago. This suggests an absence of ice from the area except for a stray iceberg twelve thousand years ago. In plain language the Ross Sea was ice-free between 13,000 BC and 6,000 BC. What caused the ice to start retreating fifteen thousand years ago? With the low sea levels in this period much of the Ross Sea was an extension of the Antarctic Continent.

Where the Azores region borders onto the Telegraph Plateau in the Atlantic Ocean the shallow seabed slopes sharply downward into the Atlantic Ocean. In the narrow strip formed by this steep slope there is an unusual proliferation of steep rock pinnacles and sharp rocks. Their shape is well preserved. There has been no rounding off of the steep pinnacles or the deep hollows that would have occurred if they had been submerged for a long time. The area had to have been above sea level less than 15,000 years ago around 13,000 BC.

Around 13,000 BC the Aleutian Islands were the partly submerged southernmost extension of the Beringian landbridge linking Asia to North America. The Aleutian Islands extend for twelve hundred miles southwest from Alaska to Kamchatka in Siberia. Much of the Bering Sea in this period was a large landlocked lake as sea levels were 425 feet lower than the present day in the early Twenty-first Century. The Aleutians would have been a more temperate route from Asia to North America.

The landmass of Beringia connected Siberia in Asia to what is now Alaska in North America. Around 13,000 BC it was a flat scrub covered windswept plain. In the twenty-first century it is the Bering Strait. The maximum extent of Beringia was around eighteen thousand years ago and it had been in existence for one hundred thousand years. With the great warming of the Bolling Interglacial the land bridge shrank at the edges before the rising seas covered the landscape.

The Russians hypothesize that Beringia was a steppe on the Siberian side that became tundra on the American side. The grassy flora of Siberia resembled that of East Africa and only started to resemble tundra on the American side. Of the thirty four species of extinct animal in Siberia twenty eight of them were from temperate climatic zones. Alaska was largely icefree whilst Siberia was completely ice free. Was Siberia more like a savannah rather than a steppe?

Around 13,000 BC the last of the great glaciers started retreating from Canada during the Bolling Interglacial.

Before 13,000 BC the mean July or summer temperature in England was ten degrees Celsius. The studying of beetle remains show that in only three hundred years, by 12,700 BC the summer temperatures rose rapidly to twenty degrees Celsius before cooling down to an average of fourteen degrees in 11,000 BC. This is ten degrees Celsius in three hundred years. This is global warming that would make your head spin. And it has occurred over and over and over.

Around 13,000 BC the European Glacial Icemass suddenly withdrew so far north that the flow of meltwater to the Euxene Lake ceased. The Euxene Lake was in the area of what is now the Black Sea and was fed by the Upper reaches of the Dniester, Dnieper, Don and Volga Rivers. The glacial mass had diverted all of these rivers west over Poland and Germany and the Euxene Lake kept retreating.

Midway between England and Denmark in the North Sea is the Dogger Bank. This was dry land in this period and fishermen trawling it have dredged up antler spear points and other artifacts. This now submerged area is now three hundred kilometers offshore. This was a totally different world to the one that we know now and it was not long to last.

In the caves of La Marche, also called Lussac-les-Chateau in Vienne in the Poitou-Charente in France were more amazing discoveries. Prehistoric stones found at Lussac in the Poitou, France, show engraved drawings of men and women dressed in completely modern style with hats, jackets and short trousers. These carvings date back fifteen thousand years and when they were dug up in 1937 the Archaeologist Stephane Lewoff remarked with astonishment that the men, women and children were shown wearing hats, shoes, trousers and skirts exactly like those of today when he wrote it, or 1937. The men wore hats, jackets, trousers and a petticoat together with boots and shoes. Are these very dressy *Neanderthals*? Or were they stylish *Cro-magnons*? Other stones depicted men and women in casual poses wearing robes, boots, belts, coats and hats. One engraving is of a young woman who appears to be sitting and watching something. She is dressed in a pantsuit with a short-sleeved jacket, a pair of small boots and a decorated hat that flops over her right ear and touches her shoulder. Resting on her lap is a square flat object that folds down the front like a modern purse. Other stones show men with clipped beards and moustaches who were wearing soled and heeled high boots similar to medieval ones. Oddly enough there are no soot marks on the walls or ceilings of the caves so the method used to illuminate them is a complete mystery unless we believe that they were executed in the dark.

More perplexing engravings were found in the Grotte De Marsoulas in Haute-Garonne in the Midi-Pyrenees in France. There is an engraving here of a horse wearing a bridle that is fifteen thousand years old around 13,000 BC. Another bridle? Another horse?

One of the most outstanding discoveries in the nineteenth century was found in Altamira near Santander in Cantabria in Spain. In 1878 after attending the Paris Exposition Don Marcellino de Sautola decided to look into a cave that his dog had gone down into in 1858. Don Marcellino had sealed up the entrance for safety reasons and now was fascinated by Ice Age tools that he had seen in Paris. When he went into the caves again he soon discovered a hand-axe and some stone arrowheads. One day his five year old daughter Marie came into the cave with him and cried out in excitement. She had seen pictures of charging bulls on the walls of part of the cave whose low ceiling had made it impossible for her father to see. Don Marcellino made his announcement to the world but was ridiculed as a fake. The paintings were still wet. Later when other cave paintings were discovered he was posthumously exhonerated by a man called Carthailac who had discovered similar paintings at Les Eyzies.

Around 13,000 BC the human population suddenly left Greece for no known reason. There had been sporadic human settlements in Epirus, Boetia and the Argive Plain since around 20,000 BC. Humans would not return to Greece until 8,000 BC. What was happening with the warmer weather? Why was the country now known as Greece suddenly abandoned in this period?

The remains of mastodons, rhinoceros, hippopotamuses and elephant have been found in the pre-glacial beds of Italy. These animals were slaughtered outright and so suddenly that few escaped. What sort of cataclysm killed all of these animals?

The Mediterranean Sea was a much smaller inland sea during the Holocene Era and supported a population on what is now sea bottom. Other sources state that the Mediterranean Sea was actually a series of large lakes. The Atlantic Ocean broke through the Straits of Gibraltar, the Pillars of Hercules, flooding and filling the Mediterranean basin around 9,000 BC to 10,000 BC according to several sources. Other reports state that this happened in 13,000

BC when the Atlantic Ocean burst through the Straits of Gibraltar and flooded the Mediterranean Basin. At this time the Mediterranean Basin was composed of two great lakes separated by the Sicilian land bridge joining North Africa and Italy. This would have drowned massive numbers of Aurignacian settlers on the lake shores. Legends of the Great World Flood could have descended from this cataclysmic event. Or is this the later drowning of the Euxene Lake which was to become the Black Sea? Or was it flooded from the Euxene Lake? The Mediterranean was largely dry land before it had been flooded from the Atlantic. Core samples taken by the "Glomar Challenger" indicate gypsum, salt and other evaporative materials. Princeton Geologist Professor Kenneth J. Hsu postulated that the Mediterranean previously was a desert. How many times has it been flooded?

During the last advance of the Pleistocene Age the Mediterranean was a great low plain with a pair of large lakes separated by the ridge connecting Italy, Sicily and Malta with Africa. During the warm interglacial periods apelike men hunted pigmy elephants and pigmy hippopotami in Sicily. With the final melting of the ice caps the ocean rose until about 15,000 years ago it broke through the Isthmus of Gibraltar and filled the Mediterranean Basin to its present level forcing thousands of men and animals to flee and flooding an enormous area. These dates are in dispute but almost all of the dates presented here are. Where are the problems with dating? Are they with the processes or the authorities? At times you will see what appear to be the same event or discovery and a different series of dates. If we cannot find a single date than I will include all of them and then see how it fits in with other data from the same period.

Sicily along with Malta was once part of a lost world of large lakes that was connected to Africa and populated with pigmy elephants and hippopotami. Dwarf elephants have been found in Sicily, Sardinia, Crete and Cyprus. The smaller ones are later than the larger ones, as they are found higher up in the stratified deposits. The remains of pigmy hippopotami were also found along with the bones of deer, bear, fox and other animals no longer found on Sicily. The fossils of the hippopotami were so common that they were mined as a source of charcoal.

You may notice that the paragraphs can be disjointed but they are all based on separate reports or studies. You can decide which ones you want to investigate or ignore.

North Africa was connected to Malta which was at the time connected to Sicily which was connected to Italy before the Great Cataclysm. Malta was connected to Sicily which was connected to Italy. Malta itself was connected to Africa basically creating a massive dividing wall across the Mediterranean.

There was also a huge fertile plain crossed by many rivers in the upper half of the Adriatic Sea which joins to the Mediterranean reaching almost 200 miles south of Venice.

Plains generally extended all along the coastlines of Spain, France, Italy, and Greece where many of the islands were joined. Maybe people had not depopulated Greece but were instead living on the coastal plains which have not been excavated yet due to now being submerged?

Joseph Prestwich, formerly Professor of Geology at Oxford University concluded that the Mediterranean islands of Corsica, Sardinia and Sicily were all completely submerged on several occasions during the rapid melting of the ice sheets at the end of the last Ice Age. The animals naturally retreated as the waters advanced deeper into the hills until they found themselves embayed. They thronged together in vast multitudes, crushing into the more accessible caves until overtaken by the waters and destroyed. Rocky debris and large rocks from the sides of hills were hurled down by the currents of water, crushing and smashing the bones. It was as if a megatsunami had pushed them all down. Maybe a megatsunami did?

The hills around Palermo in Sicily disclosed an extraordinary quantity of bones of hippopotami in complete hecatombs. Prestwich stated that they were caused by an immense flood which submerged the Mediterranean islands of Corsica, Sardinia, Sicily and Malta. The bones of numerous animals of many different types were crushed and splintered and appeared to have been jammed into caverns and fissures by a tremendous force. This was around fifteen thousand years ago.

Mezhirich near Cerkaska in the Ukraine is famed as a mammoth village. There were the remains of four huts made from 149 mammoth bones dating back 15,000 years here. A farmer, digging his cellar, almost two metres below ground level, struck the massive lower jaw of a mammoth with his spade. The jawbone was upside down, and had been inserted into the bottom of another jaw like a child's building brick. In fact, as subsequent excavation showed, a complete ring of these inverted interlocking jaws formed the solid base of a roughly built circular hut four or five meters across. About three dozen huge, curving mammoth tusks had been used as arching supports for the roof and for the porch, some of them still left in their sockets in the skulls. Separate lengths of tusks were even linked in laces by a hollow sleeve of ivory that fitted over the join. It has been estimated that the total of bones incorporated in the structure must have belonged to a minimum of ninety-five mammoths. A mammoth skull was also found here that has red painting on the front of the skull. The mammoth skull was found at the front of one of the huts. What this was for is unknown but some researchers believe that the colour red represented life or blood. A map was found here in 1966 that was engraved on a mammoth tusk. The map shows a local river flanked by a row of houses and was dated to ten thousand BC.

Alaska was unglaciated around 13,000 BC except in the area of the Alaska and Brooks Ranges. There was a continental shelf extending outwards from central Beringia along the southest shore of Alaska. The climate was dry but cold.

Onion Portage is located at a prehistoric caribou migration river crossing on the Kobuk River one hundred and twenty-five miles upstream from where the Kobuk River now enters the sea near the Bering Strait. A sandy knoll dominates the otherwise flattened area which served as a lookout for game. Human remains up to 15,000 years old have been found here. Onion Portage is in the Kobuk River Valley in Alaska.

A human habitation site dating back to 13,000 BC was discovered in Blue Fish Caves in the Yukon area of Alaska. They are only fifty kilometres southwest of Old Crow.

Among the oldest human habitations east of the Mississippi were uncovered south of Pittsburgh in Allegheny County in Pennsylvania dating back 14,000 to 15,000 years or 12,000 BC to 13,000 BC.

The first edition of the American Journal of Science, volume one, Number one, page 155, describes a discovery by a naturalist named Isaac Lee in a stretch of sandstone a quarter of a mile north of Pittsburgh on the same side of the Monongahela river. There was an unusually flat rectangular surface that was three feet long and varying from five to six inches wide. One end was cut off by a break in the rock and the other end terminated in the middle of the rock face in a straight square line as if a roll of paper had been torn off clean. On this flat surface were inscribed row after row of evenly spaced perfect diamond shapes each with an oblique raised band across its centre. Lee measured the carvings and made drawings of them and when he came back to study them again a quarry man had already removed them. Lee also took meticulous notes of the position of the rock surface in relation to the geology of the surrounding area. In fragments of the impressed rock Lee found fossils of primitive jointed plants which appeared in the Devonian Era, 400 million years ago. How do we get our dates again? Is there ever dispute about them? How exact is geology? I will just present them. You can do what you want with them.

Mount Saint Helens in Washington State erupted around this time leaving sediments of ash between layers of sediment from the Lake Missoula glacial floods. From these sediments it was worked out that there were up to one hundred gigantic Lake Missoula floods in this period that were repeatedly breaking up the mass of glacial ice.

Some reports state that thousands of cometary fragments hit Canada and the United States of North America in this period around 12,900 BC.

Sudden global warming occurred around 12,700 BC as temperatures suddenly shot up worldwide by ten degrees Celsius. This was recorded in samples of plants and insects from Greenland. This temperature rise occurred in only one year and lasted for the next fifty years.

In the Meadowcroft Rock Shelter on Cross Creek, a small tributary of the Ohio River in Pennsylvania, pre-Clovis period artifacts have been found dating to the period of 11,950 BC to 12,550 BC. Meadowcroft is near Avela in Washington County. Some radio-carbon dates suggest that Meadowcroft dates back to 17,000 BC. Items that have been found there in what some describe as the oldest paleoIndian site in North America are ceramic items, bifaces, bifacial fragments, lamellar blades, a lanceolate projectile point and chipping debris. A hearth indicating the use of fire has also been found from this period. In the Meadowcroft rockshelter near Pittsburgh James Adovasio and his team in 1973 discovered eleven distinct occupation layers dating back to 10,800 BC though burnt wood from one floor level gave a date of 13,000 BC and one very deep level containing the burnt remains of matting or basketwork dated back nineteen thousand years to 17,000 BC. This is the problem with dates. Every scientist tends to deduce their own so you end up with what appear to be the same happenings but at different dates. This is the problem with dating systems. It is a lot to do with interpretation. How reliable are our dating systems? This is why you will find events repeated at different periods in this text. I have got to stop repeating this.

Around 12,500 BC world sea levels were two hundred feet lower than the present day at the beginning of the twenty-first century. Previously around 13,000 BC sea levels were 425 feet lower than the present. This indicates a sea level rise of two hundred and twenty five feet in only five hundred years which is around six inches per year. This also indicates massive glacial melting.

Around 12,500 BC the first pulse of meltwater from this melting of the ice sheets covering northern Europe and Asia was so large that it filled and fed dozens of "great lakes" that no longer exist. These lakes were the Upper Dnieper, Upper Volga, Dvina-Pechora, Tungusta, Pur and Mansi whose combined surface area plus that of the now swollen Aral and Caspian Lakes dwarfed that of the single Euxene or Black Sea Ice Age lake by a factor of four to five. These enormous lakes filled the sag in the earth's crust caused by the weight of the huge ice mass to the north. The lakes themselves were dammed on their southern margins when the weight of the ice sheet at its maximum extent pushed the soft interior of the Earth aside. Eventually these great lakes expanded until one by one they breached the crest of the bulge dam and flowed southwards to the Aral, Caspian and Black Seas which were strongly freshened by the discharge. These were the New Euxine deposits created by the first great meltwater pulse.

Around 12,500 BC in the Middle East and Egypt the climate during the initial deglaciation of the Bolling Insterstadial was warmed by shifting patterns of atmospheric and ocean circulation and was a period of post glacial warming before the sudden climatic interruption called the Younger Dryas wherein the Ice Age suddenly returned between 10,500 BC and 9,400 BC.

Isna, also called Esna, near Qena in Egypt is one of four sites from the late Paleolithic period circa 12,500 BC that were occupied by the Isnan or Qadan peoples. The other three sites are Naquada, Tushka and Dishna in Muhafazat Aswan in Egypt. They are around 200 kilometres upriver from Aswan. These people cultivated wheat grass, wild barley, and other types of grasses. Stone sickle blades were used to reap the harvest and grinding stones enabled them to extract the maximum amount of grain. They also had mastered animal husbandry and possessed a sophisticated microblade industry, living in communal villages. Around 9,500 BC the technological skills disappeared and were replaced by crude stone implements. Agriculture itself vanished entirely from Egypt for another 4,500 years. Where did the people go?

Incidentally up to Roman times North Africa including Egypt was Rome's granary. Herodotus, the Greek historian, stated that you could travel from Egypt along the coast to Morocco and you would see the same forests that you travelled through all the way. Roman overcropping and salination followed by Arab goat raising practices turned this once lush and fertile area into the desert that it is today. Don't let us assume that we were the first population to start destroying the environment on a massive scale. It seems to be a historic fact that civilization is always followed by ecological degradation. Anthropogenic or man-made ecological degradation started in Mesopotamia and in fact wherever man first established a community as the centre of a culture. Wherever a culture started there was ecological degradation. Quite often desertification followed. This in many cases was merely due to the quest for wood for charcoal for cooking in these new population centres. No one thought ahead.

Around 12,300 BC the Great Lake Bonneville Flood occurred. In only a matter of a few days it carved out the Snake River Cayon near Twin Falls in Idaho. Estimates are that fifteen million cubic feet of water per second charged through here. At Portneuf Narrows the water was four hundred feet deep. This was northwest of Redrock Pass. Lake Bonneville had covered much of Idaho, Nevada and Utah. The lake was one thousand feet deep in parts and more than 51,000 square kilometres in area. When the Great Lake Bonneville Flood occurred it charged initially through the Redrock Pass. All that remained of Lake Bonneville after the great flood were Great Salt Lake, Utah Lake, Sevier Lake, Rush Lake and Little Salt Lake as remnants.

The tally of large meteorites to strike the earth in the Thirteenth Millenium is nil apart from some reports that state that thousands of cometary fragments hit Canada and the United States of North America in this period. Maybe there were craters left in the seas. This is the period when our dates might not be so reliable after all as this same phenomenon of impactoid fragments hitting the United States and Canada was reported only a few thousand years later. Did we have several separate impactoid events onto the same area in only a small period of time or were the dates mixed up? A lot of our researchers as you may or may not have observed so far into the text leave a wide mark of one thousand years for an event to occur. This gives dating a very broad latitude.

12th Millenium BC.

During the 12th Millenium world sea levels were one hundred metres or three hundred and thirty feet below present sea levels. This was one hundred and thirty feet lower than only five hundred years before. Glaciation was indeed rapidly increasing again. There was another Ice Age on the way again.

On March 1, 1976 it was reported by the world press that an Italian expedition had discovered a now petrified forest 250 metres wide by 2,000 metres long under the ice of Antarctica. This proved that 14,000 to 16,000 years ago Antarctica enjoyed a moderate climate. How could this be? You know. This could be only if the actual planet Earth had tilted suddenly after this period, leaving the climatic zones intact and still in their old positions but with new lands positioned under them. As you will see there actually was a planetary tilt in this period.

Hapgood concluded from the finds of tropical fossils in the area that there was a period when the Antarctic continent was on the Equator and then driven away from it. Between ten and fifteen thousand years ago, the Antarctic was 2,500 miles further north. The climate was mild. Suddenly an ice age began with the ice first accumulating at the pole then melting and reaching the temperate zones. Owing to the centrifugal force produced by the two centres of gravity of the Polar Caps the Earth's crust began to slide causing Hudson Bay and Quebec to be displaced 2,500 miles southwards, Siberia northwards and the Antarctic southwards. In a few thousand years, the Antarctic reached the South Pole. This was Hapgood's hypothesis. The shift occurred but it might not have been by crustal displacement.

An asteroid or comet of huge dimensions could have done it much more efficiently and much faster. Think of billiard balls. You just have to kiss one with the cue and amazing things happen.

The Russian hydrologist M. Ermolaev showed that the present Arctic water system was established around 12,000 years ago, the date being another ending of the glacial epoch in Europe and North America.

The glaciers of the Ice Age covered the greater part of North America and Europe while the north of Asia remained free. In America, the plateau of ice stretched from latitude 40 degrees north and even passed this line. In Europe, it reached latitude 50 degrees north. In northeastern Siberia, above the Polar Circle, above latitude 75 degrees, there was no perennial ice.

Hapgood stated that Hudson Bay in Canada was the North Pole during the end of the last Ice Age with the North American Ice Cap covering 4,000,000 square miles. Hapgood also states that there was a shift in the Earth's crust that started 17,000 years ago and took 5,000 years to complete. During this shift North America and the Western Hemisphere was moved southwards whilst the Eastern Hemisphere was moved northwards. The great North American Ice Cap was melted suddenly and Siberia was then placed in deep freeze. The South Pole would have then been located off the Wilkes Coast and the whole continent would have been ice-free.

There are various theories about the time taken for the Earth to tilt but to actually snap freeze mammoths and other animals it would have to be quite sudden as the climatic zone over them changed from temperate to Arctic quite suddenly. Otherwise the bodies would start to putrefy. We could be talking a very short time period indeed and not even in geological terms.

Examples of coral offshore from Newfoundland in Canada indicated that a mini-Ice Age that developed over only five years occurred in this period.

Lake Livingstone which was a vast lake in North America suddenly collapsed releasing 84,000 cubic kilometers of water and raising sealevels at least nine inches. This was followed by the return of the cold again due to the influx of cold water into the Atlantic Ocean which upset weather patterns.

Some researchers suggest that the St Lawrence River appeared at this time diverting the Canadian glacial waters from running into the Mississippi and instead running into the St Lawrence River.

Around 12,000 BC at Glozel near Auvergne in Vichy in France in 1924 or 1921 bricks, axes, pottery and tablets of the Magdalenian period were found by a sixteen-year-old boy named Emile Fradin. One incised tablet in particular shows a collection of signs or letters that are equivalent to Phoenician or Greek while others are unidentifiable. The Glozellian tablets date from thousands of years before the Egyptians developed hieroglyphic scripts and well before the Greeks. Hundreds of tablets, engraved stones and potshards with characters on them have since been found. In 1928 a commission of Swiss and French authorities marched across virgin soil that Emile Fradin, the discoverer of the stones, had not tilled. The remains of bones were found that dated to twelve thousand BC. There were also several thousand interesting stones and some clay tablets as well as urn-like vessels that contained alphabetical as well as mathematical symbols. The finds at Glozel date from ten to fifteen thousand years ago. Bricks, inscribed stones, two paring knives, two small axes, and two rocks bearing inscriptions as well as curious vessels that look like skulls clad in space helmets. There are pottery phallic symbols and animals inscribed on bones and pebbles. There were even inscriptions of reindeers and panthers which had been extinct in the area for at least ten thousand years. The

Glozellians were quite capable artists as a carved stone horse has been found here that is exceptionally realistic and could not be misidentified as anything else except as a primitive and early wild horse. The authorities declared the objects to be fraudulent. Forty-five years later Mr Gavn Majdal, a physicist on the staff of Denmark's Atomic Energy Commission, used thermoluminescence dating on the clay objects after being approached by a Mr Sture Eilow, a Swedish amateur archeologist. The tests proved conclusively that the Glozel artifacts were actually from the Magdalenian period.

By 12,000 BC birch forests that had previously not existed now covered much of England and much of western and northern Europe. Other trees also began to occupy what were previously grasslands and tundra. This was the middle of the Bolling Interglacial before the Younger Dryas reverted back temporarily to Ice Age conditions. Birch forests need a warmer climate compared to pines.

Around 12,000 BC beech or *nothofagus* forests started covering New Zealand in the South Pacific Ocean as the temperatures rose with the retreat of glaciation. The previous open vegetation and grasslands that were here during the Ice Age were gone in only three hundred years. Where had the beech trees come from?

Around 12,000 BC the Scandinavian ice sheet rapidly melted during the great warming. This melting released billions of litres of freshwater into the oceans and sea levels were rising up to forty millimeters a year. Whilst this melting occurred birch forests rapidly covered Scandinavia and northern Russia whilst the tundra and steppe effectively vanished.

Around 12,000 BC the Alpine Ice sheet in Switzerland rapidly melted during the great warming. This melting released billions of litres of freshwater into the oceans and sea levels were rising up to forty millimetres a year.

All over North America the fluted Clovis Point spearhead suddenly appears from Washington State in the northwest to Mexico in the south and then across to Florida in the southeast. None are found in Alaska so in which direction did they come from? If they were said to have arrived via Alaska then why is there no evidence of them travelling through. There were none found in Siberia either.

Near the Aucilla River in Jefferson County in Florida amongst numerous remains from this period of humans and mammoth and mastodon bones a carved mastodon ivory tusk was discovered along with stone tools. The Aucilla River flows underground in a lot of places. The area that the river flows through is riddled with limestone caves.

Finally we get to Folsom Man discovered at Folsom in Union County in New Mexico. These are the discoveries of Folsom Points or Folsom Projectiles, which were found near the remains of Folsom Man who lived 13,000 to 14,000 years ago. They left a deep wound in the form of a neat slot. They are of stone, only five centimetres or two inches long and similar to a lance shaped leaf. The distinguishing feature of the Folsom Points is characteristic fluting that may have been used to help in hafting to a spear shaft or dart. They were not arrowheads or lance points or javelin tips yet they were motivated with exceptional force of penetration and used for hunting big game. How they were fired is unknown. Folsom Man in this area was hunting the now extinct straight horned Taylor's Bison with his peculiar spearheads grooved on the sides like a bayonet. The remains of Folsom man were discovered in 1926 in Wild Horse Gulch near Folsom. A flint that had been artificially shaped into a spearpoint was discovered deep in clay alongside the bones of an Ice Age bison. Later that same year another flint spear was found wedged in the ribs of another Ice Age bison. This spear point in the second bison dated to 9,500 BC.

Lake Lahontan, a glacial lake, was at its maximum extent across Northern Utah and Nevada. During this period the lake covered 8,500 square miles or 22,000 kilometres. What is called the Carson Sink would have been at its centre and the lake had an average depth of nine hundred feet where Pyramid Lake is today.

There was a minor global warming period around 11,700 BC that lasted fifty years. Temperatures rose ten degrees Celsius in a space of only two years and this temperature variation lasted for the next fifty years. This had previously happened one thousand years before and was evidenced by the remains of plants and insects in Greenland that are now covered by two kilometres of ice. What was the cause of these massive temperature increases?

Now we head to Lapa Vermelha near Confins near Lagoa Santa in Minas Gerais State in Brazil. On October 1999 scientists in Brazil examined a female hominoid skull that had been found in 1975 at Lapa Vermelha. This skull was named Luzia and the scientists dated her death to 11,500 BC. Her features were reconstructed in England and she appeared African rather than Amerindian. Luzia was not an Asian. Incidentally the shortest distance between the Old World and the New World is the short space between the west coast of Africa and the east coast of Brazil. Luzia's people arrived in the area around 13,000 BC. Sea currents travelling from Africa to South America would have allowed even unguided craft such as rafts to travel easily from one continent to the other. Were Luzia and her people intentional or accidental visitors to South America? The extremely low sea levels would have helped even further. The distance between Africa and South America, the Brazilian coast, is only three thousand kilometres now. With the drop in sea levels then it was much smaller.

At Monte Verde on Chinchihuapi Creek northwest of Puerto Montt in south central Chile more remnants of ancient man were found which were originally comprising two sites a few hundred feet apart that had been flooded under peat and preserved. Hundreds of stone, bone and wooden tools, animal bones, even mastodon bones and flesh, cylindrical spear points, timber poles with attached pieces of knotted reeds that were the remains of hide covered huts, as well as remains of firepits and a childs footprint were found in undisturbed layers so finding relationships and sequences was simple to establish. This was a site of human habitation 12,500 to 13,500 years ago. The site is located in a boggy area in which perishable plant and animal matter was well preserved. Two pebble tools were found hafted to wooden planks. Twelve architectural foundations made of cut wooden planks and small tree trunks staked in place were also found, as were large communal hearths as well as small charcoal ovens lined with clay. Some of the stored clay bore the footprint of a child eight to ten years old. Three crude wooden mortars were found, held in place by wooden stakes. Grinding stones were uncovered along with the remains of wild potatoes, medicinal plants and seacoast plants with a high salt content. These types of seaweed were used in medical compounds. Could this suggest oceanic origins for the people here? From where had they come? There were the remains of domestic amenities. At the oldest level there was found a split basalt pebble which is a kind of primitive tool, some wood fragments, two modified stones and some charcoal dated at 33,000 years old around 31,000 BC.

The priests of Heliopolis in Egypt told Diodorus Siculus that the primeval chaos of the first Gods was a Deluge, the same as that of Deucalion of Greece. The Gods had given shape to chaos out of their divine will. Out of this chaos they had formed and populated the land of Egypt wherein for thousands of years they had ruled among men as divine Pharaohs. The priest said that Egypt was where the first genesis of living things was and that the inhabitants of southern Egypt had a better chance of survival than any others. Diodorus was told that this was due to its geographical situation. When the moisture from the abundant rains which fell among other peoples was mingled with the intense heat of Egypt the air became very well tempered for the first generation of all things. This is corroborated by evidence indicating that when the last Ice Age was ending and millions of miles of glaciation were ending in Europe at the same time the rising sea levels were flooding coastal plains. When the huge volume of extra moisture was being dumped as rain Egypt benefited for several thousand years from an exceptionally humid and fertile climate. This started around 11,500 BC.

Shortly before thirteen thousand BC grinding stones and sickle blades with a glossy sheen on their bits, the result of silica from cut stems adhering to a sickle's cutting edge, appear in late Paleolithic toolkits in Egypt. These grinding stones were used in preparing plant food. At the same time fish stopped being a significant food source as evidenced by the absence of fish remains. The replacement food resource is ground grain, namely barley. The pollen profile of barley appears just before the first settlements were established. As suddenly as the protoagriculture appeared it vanished. After 10,500 BC the early sickle blades and grinding are replaced throughout Egypt by Epipaleolithic hunting, fishing and gathering peoples who used stone tools. The civilization had effectively relapsed. This was after unprecedented Nile floods during the eleventh millennium BC. Was this backwash up the Nile Valley as the waters from the breach of the Pillars of Hercules at the Western end of the Mediterranean to the east where they would later break through the Bosporus in Turkey thus flooding the Black Sea? Or was the breach of the Euxine Lake, now the Black Sea, much later, in fact in 5,600 BC? Was this backwash then the flooding of the Mediterranean Basin? There has been controversy over when the Mediterranean was flooded and one thousand five hundred years is not a long time geologically speaking.

Between 12,500 BC and 9,500 BC there were communities along the Nile River that had advanced tool making technology as well as domesticated animals as well as early agricultural techniques. Then they disappeared.

Around 11,500 BC the Sahara, a relatively young desert was green savannah until the tenth millennium BC. This savannah was brightened by lakes boiling with game and extended across much of Upper Egypt. The delta area further north was marshy but dotted with many large and fertile islands. The climate was significantly cooler, cloudier and rainier than today. For around 2,000 to 3,000 years around 10,500 BC it was almost continuous rainfall. Then the floods came. After the floods the arid conditions set in. This period of desiccation lasted until 7,000 BC when the Neolithic subpluvial began with one thousand years of heavy rains followed by 3,000 years of moderate rainfall which proved ideal for agriculture. For a long time the Sahara Desert bloomed and supported communities that it cannot do today.

Under the Sahara desert is the vast underground sea called Albienne, covering 230,000 square miles. Is this where the water drained that fell in this period other than that which evaporated as the desertification occurred?

The remains of a community dating back to 11,500 BC were found in Tell Abu Hureyra in Syria. It was composed of numerous dwellings that were dug partially in the ground then roofed with branches and reeds supported by wooden posts. The site was occupied for generations and the remains of 712 types of seeds were found in the occupation layers indicating massive roaming for plants. Forests of oak and pistachio and other nut trees were within easy walking distance. Now the nearest forest is at least one hundred and twenty kilometers away. Pistachio trees once grew in long lines on low wadi terraces only a short distance from the village. This is very advanced agriculture for this period. The inhabitants also had access to two forms of rye as well as wheat. Today these grasses do not grow within one hundred kilometers from here. This farmer's village was on the Upper Euphrates River in Northern Syria. There is evidence of cereal cultivation here from the discovery of stone pestles, rubbing stones and milling stones. Also found in abundance were three types of cereal grains that are not indigenous to the area. These are a form of wild barley, wild wheat called einkorn and wild rye. Two of these had been grown by Paleolithic communities such as Isnan on the Nile River in Egypt. The plants must have come to the area from somewhere and it is unusual that the rise of the cereals in the area is after 9,500 BC and the Egyptian communities decline starts at the same time. The two areas are 900 kilometres apart. Were these people actually Egyptian refugees? The earliest use of cosmetics was found here as well. A large cockleshell dating from 7,000 BC was found to contain traces of powdered malachite, a green natural substance known to have been used as eye shadow in predynastic Egypt. Malachite is not native to the region. Extremely long beads made of hard stones, seven on the Mohr hardness scale, have been found here. These are of agate, carnelian and quartz and were up to 5.5 centimetres in length. We need highly specialized diamond tipped tungsten carbide drills to cut these and the drills have to be constantly cooled by running water. The beads date to between 7,500 BC and 7,000 BC. How they were drilled is still a mystery.These dates indicate a very long occupation period at this location.

There was sudden and widespread human settlement in Alaska around 11,500 BC during the period of warming. Had these early people come from Beringia or Siberia? Or had they arrived in North America via the Aleutians?

Now we finally arrive at Clovis in Blackwater Draw in Curry County in New Mexico. From 1932 onwards projectiles similar to the Folsom ones were found on the border between Texas and New Mexico and west to Naco in Arizona. They are from ten to twelve centimetres long, 4.8 inches, and date back at least to 8,000 BC to 11,000 BC. They were able to penetrate the skull bone of a mammoth. First found in 1932 these thin flint blades were first discovered in Blackwater Draw near Clovis by a highway construction company. These were fluted flint spear points that were larger and longer and in fact earlier than the Folsom spearpoints. Some scientists reckon that these remains actually date back to 11,500 BC. Human remains were later found in 1936 and 1937 that dated back to 9,000 BC.

The tally of large meteorites to strike the earth in the Twelth Millenium is nil.

11th Millenium BC.

There are a lot of controversies around the end of the last Ice Age.
When did events actually occur?
When did it actually end?
There are many disparate dates from 11,000 BC to 8,000 BC. How do we get a truer picture of this period? In these one thousand year periods many events occurred that might not even be correctly dated due to fluctuations in radiation levels or oxygen isotope levels. We might have to combine them together?

The Astronomer Fred Hoyle proposed that only 13,000 years ago around 11,000 BC, New York was covered with several hundred metres of ice as it had been for the preceding one hundred thousand years. Suddenly the glaciers retreated all over Scandinavia and North America. In Britain, the temperature shot up from a summertime value of only eight degrees Celsius to 18 degrees Celsius in only a few decades. The temperature though began to fall again until 11,000 years ago around 9,000 BC, when the glaciers were back but not to their previous extent. In Northern Britain they covered mountaintops but not valley bottoms as they had before. Then about ten thousand years ago, 8,000 BC, a second warm pulse occurred. The temperature shot up by ten degrees Celsius in only a geologic moment. Thus ended the Ice Age that had lasted one hundred thousand years. This temperature increase occurred within one human lifetime. Hoyle postulated that something had to have washed out high-level atmospheric haze thereby increasing the water-vapour greenhouse sufficiently to send the temperature up almost instantly by ten degrees Celsius. There also had to be a change from cold ocean to warm ocean as well in the Atlantic. This would normally be produced by ten years of sunlight with normal greenhouse effects. Water had to be thrown up into the atmosphere and stayed there for ten years. Over one hundred million million tons of water was needed. Only a comet splashing down into a major ocean could do this. Could a mega-comet have deposited masses of water-ice which then melted onto the Earth?

With the radioactive output of these cataclysmic collisions how accurate indeed are our radiation and radiocarbon readings? If these are erroneous, then our dates for many events, such as mass extinctions and geological uplifts, may be out as well.

Clube and Napier theorize that a giant comet settled into an earth-crossing orbit 50,000 years ago and for 30,000 years remained intact. 20,000 years ago, 18,000 BC, a massive fragmentation event occurred and from 17,000 years ago, 15,000 BC, multimegaton fragments may have periodically collided with the Earth thus causing a reduction in glaciation. There were two large impacts in the eleventh Millennium BC, 11,000 BC, and the ninth Millennium BC, 9,000 BC, which raised global temperatures so much that the Ice Age was brought decisively to an end.

Visualize an orange floating in space or in the air near you. If it were a planet its top parts would be the coldest parts such as the North and South Poles are on our planet. This is purely because of position. Cold areas are always on the top and the bottom and warm areas are always around the middle as on the Equator. These weather zones cannot move. Now visualize the orange itself suddenly tilting down. Vast ice masses that were at the northern and southern parts of the planet are now in warmer climatic zones as the weather conditions cannot move as the planet tilts. Being warmed the ice masses now start melting whereas the temperate savannahs that have been thrust up into the Polar Regions have now been hit by instantaneous extreme cold. The glaciation will melt faster than it will take to form at the new Polar positions. This would be a Cataclysm! You would have snap frozen mammoths. Many of them still with buttercups in their mouths! Buttercups don't grow in the snow or the Arctic. The buttercups were found in their mouths and other temperate plants were found in their stomachs!

Geologists have discovered that the hard outer crust of our Earth floats on a molten mantle and that continents rest on separate tectonic plates. There are direct relationships between the sliding of the plates and the changing of the Polar Ice Caps. The Poles seem to stay stationary for about thirty thousand years and then shift for six thousand years and then stay put for another thirty thousand years. The last four rounds of the poles started one hundred and twenty thousand years ago when the North Pole installed itself in the Yukon Territory in Canada at 63 degrees north and 135 degrees west. Then it went to the Greenland Sea at 72 degrees north and 10 degrees east about 84,000 years ago. It then moved from 54,000 till 48,000 years ago and settled in the middle of the Hudson Bay area at 60 degrees north and 83 degrees west. It rested there for 30,000 years and then wandered from about 18,000 to about 12,000 years ago to its present location. Simultaneously the South Pole performed similar gyrations but in the opposite direction. The three moves prior to the last one were all in the Southern Indian Ocean and only the last one twelve thousand years ago ended up in Antarctica. At least half of Antarctica was ice-free for the last one hundred thousand years while Palmer Peninsula and Cape Horn, which may have not been separated from each other, enjoyed a warm climate. During the preceding period Cape Horn in South America was connected to Antarctica due to reduced sea levels caused by the glaciation.

Mountain ranges were raised in some places up to six thousand feet whilst other areas collapsed. The bed of the Atlantic Ocean dropped nine thousand feet changing from a continental mass to a series of mountainous islands taking the remains of the continental mass with it and causing massive tidal waves or tsunamis across the face of the Earth.

A tremendous explosion, the type that hopefully we shall never see again, hurled rocks found only in Labrador in Canada onto mountains in New Hampshire, Massachusetts and Connecticut to the southeast and Wisconsin to the southwest which are in geographical terms the same distance away from Labrador. Was this an impact explosion like that caused by a meteor or an ejective explosion caused by volcanic activity? Did we have actual giant meteorites as well as ice comets crashing into the Earth at this time?

There is another explosion site like that found in Labrador in Canada. In the southeast United States, the Gulf of Mexico and in northeast South America there are curious phenomena called bays that are parabolic impact craters of gigantic size. Unlike the Carolina Bays these different bays contain rock only found in the Mid-Atlantic Shelf. The rocks arrived at incredibly high speed but at very low angles of trajectory thereby ruling out meteors and other related cosmic visitors. This indicates that due to the fact that these bays are on the circumference of a circle, with the strange Mid-Atlantic site at its epicenter, then indeed another eruption of horrific magnitude took place. These rocks though are not volcanic ejections but appear to be normal rock that has been blasted out of its normal position. So we have cosmic chaos as during a meteor shower of the type that the world may never have seen we have two single enormous magnitude meteor or comet collisions, one in Labrador and the other in the Mid-Atlantic Ridge, all in all enough to shift the angle of the planet tilt. Like playing billiards really.How correct is this date though?

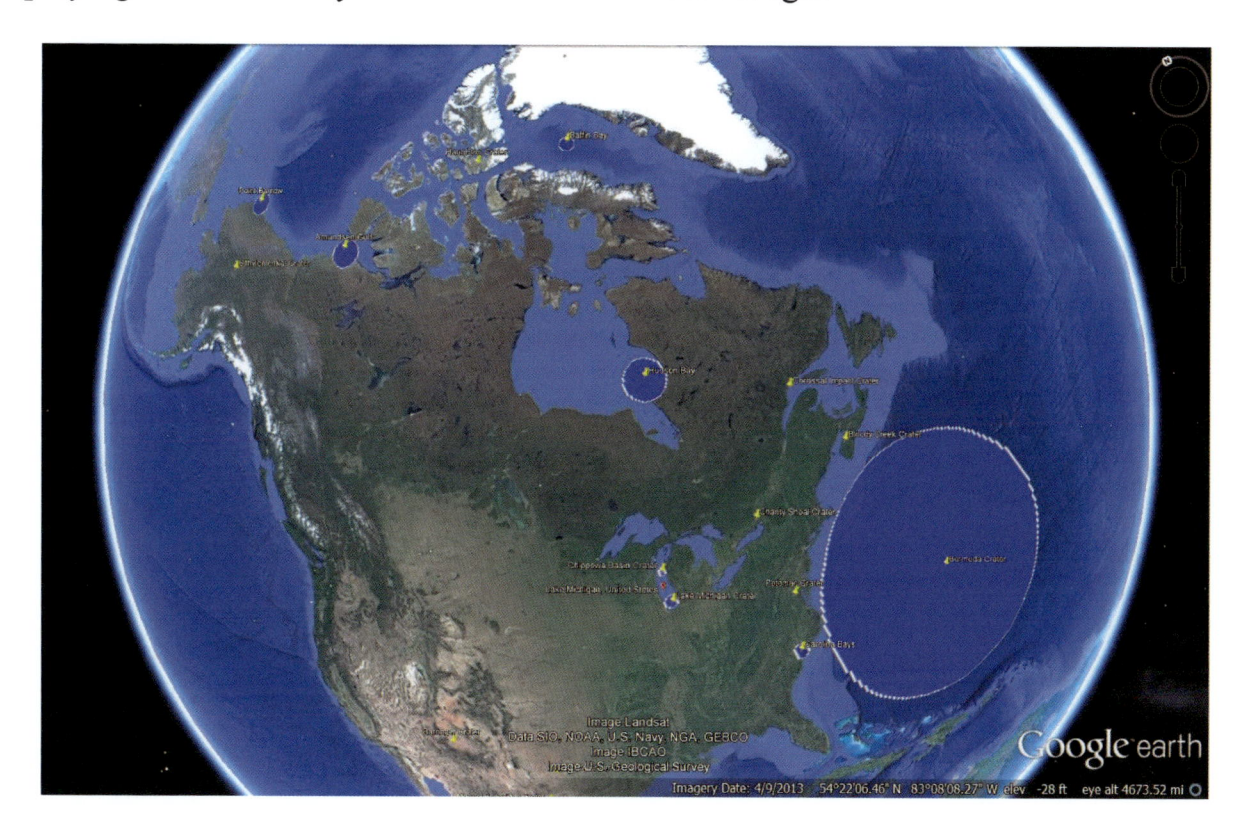

How far was matter ejected from the impacts during this three thousand year period? The large radius on the right is the Bermuda Crater, the hypothesis being that the island of Bermuda is actually the central uplift of an enormous meteoric impact. Were the three smaller impacts that are heading northwest created before or after the Bermuda impact? The earth turns to the left so they are possibly secondary impacts created by smaller meteorites that travelled in the main swarm. Or large fragments of the main impactor? This is a tear drop impact series.

Kelly and Dachille proposed that the island of Bermuda is the central peak of a gigantic impact crater in what is now the Atlantic Ocean. This crater has a diameter of 2,250 kilometres when a cosmic object struck the Earth twelve thousand years ago. This impact was responsible for massive changes to the Earth at this time and Kelly and Dachille also propose that the Carolina Bays were created by the ejecta from the impact. Other sources state that the Carolina Bays were created by ejecta from the meteoric impact into the Hudson Bay area.

Other sources state that the bays were created by cometary impacts therefore creating parabolic craters as the ejecta crashed in at low angles of trajectory.

At the peak of the last Ice Age from around 24,000 BC to 14,000 BC, so much water was locked up in the ice caps that worldwide the sea surface had fallen by an average of four hundred feet. By the end of the Ice Age, the sea levels had risen to their previous levels and the present shorelines indicating a sudden rise of four hundred feet.

What had we mentioned about the consequences of our Earth tilt?

During the late Ice Age sections of the Earth's crust previously pressed down into the Asthenosphere of the planet by billions of tons of ice would have been liberated by the thaw and begun to rise again, sometimes rapidly, causing devastating earthquakes and filling the air with terrible noise.

Europe during the Ice Ages had a narrow strip of fertile land between the Alpine glaciers in the south and the gigantic ice cap in the north which occasionally covered the whole region down to 52 degrees north. The land was sodden with melted ice and water collected in lakes and ponds and accumulated on the moors. It resembled the Siberian tundra. It was a good hunting ground but unsuitable for farmers and gardeners. Suddenly this all changed. Siberia became the new tundra. Europe deglaciated and dried out.

In 20,000 BC Europe was in mid Ice Age. Ten thousand years later it was still bitterly cold. Suddenly there was a climatic turnaround, which had led by 5,000 BC to a climactic optimum of brilliantly mild to warm weather causing all of Europe to be covered in a light green blanket of oak and beech trees over a period of one thousand to two thousand years.

The Ice Age was a global phenomenon affecting both the northern and southern hemispheres of the Earth. Similar climate and other conditions prevailed in many parts of the World notably in eastern Asia, Australia, New Zealand, and South America. There was massive glaciation in Europe, where the ice reached outward from Scandinavia and Scotland to cover most of Great Britain, Denmark, Poland, Russia, large parts of Germany, all of Switzerland, and big chunks of Austria, Italy and France. This is known technically as the Wurm Glaciation and started 70,000 years ago, a little later than in America but attained its maximum extent at the same time as the American glaciation, namely 17,000 years ago when they both experienced the same unexplained rapid withdrawal and shared the same terminal date. It then took 7,000 years for deglaciation to occur. Other sources state that deglaciation was almost immediate. Catastrophism verus Gradualism? Take your pick. Or was it both?

Chatelain argues that the Ice Ages are caused by the precession of the Equinoxes whose duration is 26,000 years and the cycle of the variation of the eccentricity of the Earth's orbit around the Sun whose duration is 104,000 years combined causes the changes of temperature and humidity on our Planet. This is 21,000 years. The fourth cycle is that of the variable obliquity of the Earth's rotational axis in relation to the ecliptic and its duration of 42,000 years. The

Glacial cycles repeat themselves every 126,000 years or so with a shorter warm period in between of about 42,000 years in between the two severest periods of ice and then a longer and warmer period of 84,000 years with a slightly colder period in the middle. It would take five periods or 630 thousand years for the cycle to be repeated.

The first consequence of deglaciation was a precipitous rise in sea levels perhaps by as much as 350 feet. Islands and land bridges disappeared and vast sections of low-lying continental coastline were submerged. From time to time great tidal waves rose up to engulf higher land as well.

There was widespread volcanic activity around the Azores Islands and the Canary Islands in the Atlantic Ocean around 11,000 BC. The seafloor around the Azores was suddenly covered with lava. The Azores are on the eastern slope of the Mid-Atlantic Ridge. The ocean basins around the ridge are only thinly coated with sediments indicating that they are very young geologically. Tachylite lava has been found here which normally disintegrates in sea water in around fifteen thousand years. This is on the seabed around the Azores indicating that the volcanic outpourings must have occurred here less than thirteen thousand years ago. Some speculate that there was an Atlantic plateau that extended westwards of the Anti-Atlas Mountains in North Africa to the Canary Islands. In many respects we have a world very different to that of today. The seas were smaller as the continental shelves were dry land before they were suddenly flooded with the sudden glacial melting.

Fairbanks studied coral reefs off Barbados in the Caribbean Sea that had been growing there for twenty thousand years. Fairbanks noticed that one particular type of coral grows only in shallow water and dies if the water gets too deep. Cores through the coral layers showed sudden changes in sea levels in this period indicating that the ice masses had melted quite rapidly and not gradually as originally theorized. There were two major jumps in radiocarbon levels around 11,000 BC and 15,000 BC. Fairbanks insists that this was a sealevel rise of fifty feet in a few weeks. This is the same as cores taken from Tahiti and the Sunda Strait that also indicate 11,000 BC. What caused world sea levels to suddenly rise fifty feet in only a few weeks? This was a sea level rise and not tsunami or tidal wave action which also occurred at this time. Was there rapid deglaciation followed by sudden sea level rises?

The Charity Shoal Crater is a shallow rimmed impact crater in Lake Ontario near Wolfe Island. The majority of the crater is in Canada and it is one thousand metres wide. The crater was created thirteen thousand years ago.

In the period of 11,000 BC the Nile River in Egypt flowed six metres higher than it does today and was narrower and shallower than the present river. It also flowed into a large landlocked lake that eventually became the eastern Mediterranean Sea.

Now we are finding a report of an impact crater from this period in Kyzilorda Oblast in Kazakhstan. The Aral or Aralsk Impact Crater is four hundred and fifty kilometres across and around thirteen thousand years old. This would help explain the origin of the Aral Sea in its basin.

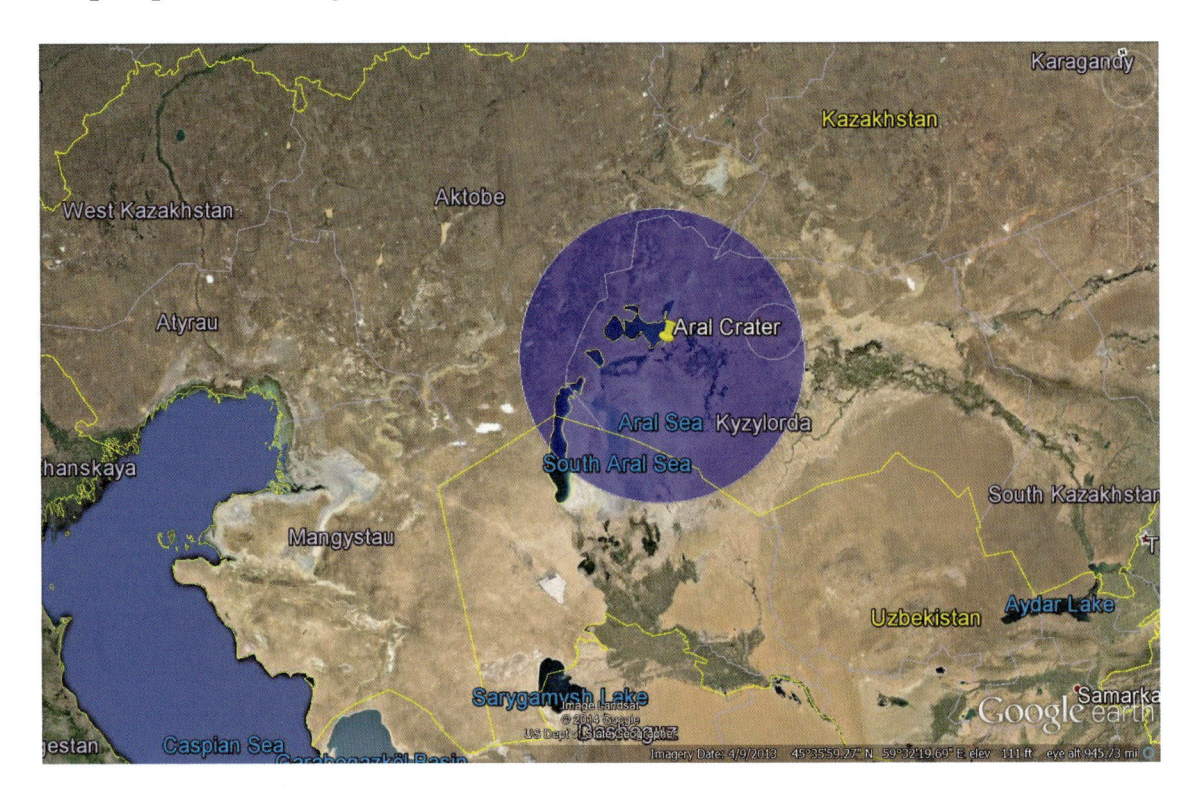

The Aral Crater. This impact could have started an Earth tilt thirteen thousand years ago. We do have later contenders though. Did the effect of all of these impacts from the east cause the earth tilt that changed the world? Three thousand years later there were would be two more major impacts in this area. Is this too much for coincidence and statistics? Or were all three impacts around the same time?

Are we seeing an increase in meteorite impacts?
Around 11,000 BC there was a major change to the previous two thousand year high rainfall period. A long period of severe drought occurred that caused major change to the increasing urbanization of the Middle Eastern communities such as Abu Hureyra with the sudden drop in crop yields and increasing of forage areas. A result of renewed glacial conditions in the north and the shutdown of Atlantic Ocean circulation with the stopping of the Gulf Stream that resulted due to harsh anticyclonic conditions caused severe drought in Anatolia in Turkey and the rest of the Levant and Middle East that lasted one thousand years.

The body of a woman, physiologically more like the Ainu of Northern Japan than the Indians of North America, was found in Mexico City. Had she or her forebears arrived on the Japan Current from across the Pacific? And on what sort of watercraft?

The Yucatan in Mexico which was warm and wet for the previous fifteen hundred years suddenly turned cool and dry. The cenotes or caves in the limestone under the surface of the ground were dry though they contained drinking water. These cenotes were inhabited. The cenotes are huge circular sinkholes going down into the limestone shelf. Are some of these caused by meteorites as well? Are the cenotes a mixture of natural sinkholes as well as meteorically induced ones? After all the Chicxulub Crater, apparently responsible for the extinction of the dinosaurs in 65,000,000 BC is in this same area. There is a higher density of circular alignments of cenotes overlaying the craters rim.

Around 11,000 BC there was a major change to the previous two thousand year high rainfall period. A long period of severe drought occurred that caused major change to the increasing urbanization of the Middle Eastern communities such as Abu Hureyra with the sudden drop in crop yields and increasing of forage areas. A result of renewed glacial conditions in the north and the shutdown of Atlantic Ocean circulation with the stopping of the Gulf Stream that resulted due to harsh anticyclonic conditions caused severe drought in Anatolia in Turkey and the rest of the Levant and Middle East that lasted one thousand years.

A map of the Heavens was found in a cave in Bohistan in the Himalayan Foothills of Pakistan. Astronomers who studied the map also noted that it diverged from our own maps in that it showed the position of the heavenly bodies 13,000 years ago around 11,000 BC, which confirmed its accuracy. Another curious feature of the map, which was published in the National Geographic Magazine, was that lines were drawn on it connecting the Earth with Venus.

Lake Lauricocha is near the Bolivian border with Peru. The Peruvian engineer Augusto Cardich discovered a culture very high up in the Andes near Lake Lauricocha that is thirteen thousand years old. In a cave the remains of human skeletons have been found that are dolichocephalic, or elongated in skull formation dating to at least 7,500 BC.

Please remember that my books are designed as gateways for you to go off and use the internet to explore at your leisure. Google Away!

And it is not alone!

Suddenly the heavens fell to Earth.

A massive meteoric storm was occurring!

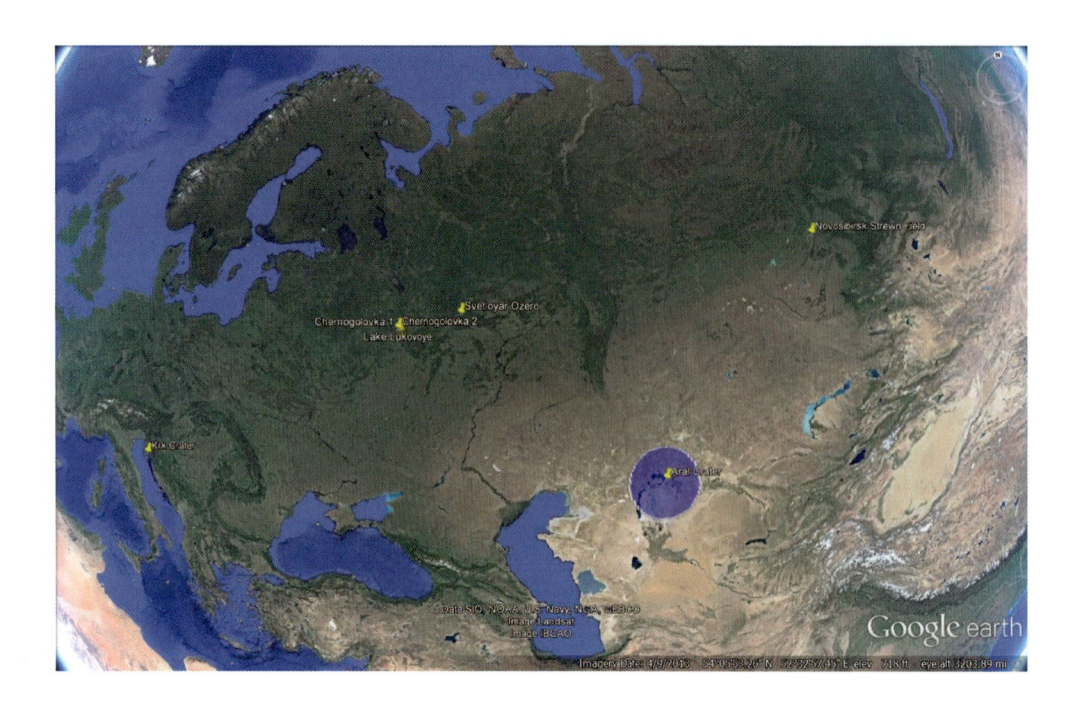

An overview of impacts in Europe and Asia. The Krk crater is to the west. The large circle is the Aral Crater. To the right is the Novosibirsk Crater. In the centre are Russian impact craters that are shown better in the next map.

There is an impact crater in the Gorkiy Region in Russia. The Svetloyar Ozero in East Nizhny Novgorod is an impact crater two hundred and ten metres long and one hundred and seventy-five metres wide. It was created thirteen thousand years ago.

The Borovoye Ozero near Moscow is an impact crater four hundred and seventy metres across. It was also created thirteen thousand years ago.

Chernogolvka 1 and Chernogolovka 2 are meteoric impact craters in the Moscow Region in Russia. Number one is three hundred metres wide and number two is fifty metres wide. They were also formed thirteen thousand years ago.

Lake Lukovoye is formed by a thirteen thousand year old impact crater that is six hundred and forty metres wide in the Moscow area in Russia. Around 10,900 BC there were massive impacts in Russia. Was this impact part of this meteorite or asteroid shower? One hundred years is nothing in geology.

The Orlevo Ozero impact crater is around thirteen thousand years old and is three hundred metres in diameter. The Orlevo Crater is also in the Moscow region.

The Svetloe Ozero near Moscow is an impact crater three hundred metres across. It was created thirteen thousand years ago.

What sort of meteoric storm was occurring in this period?

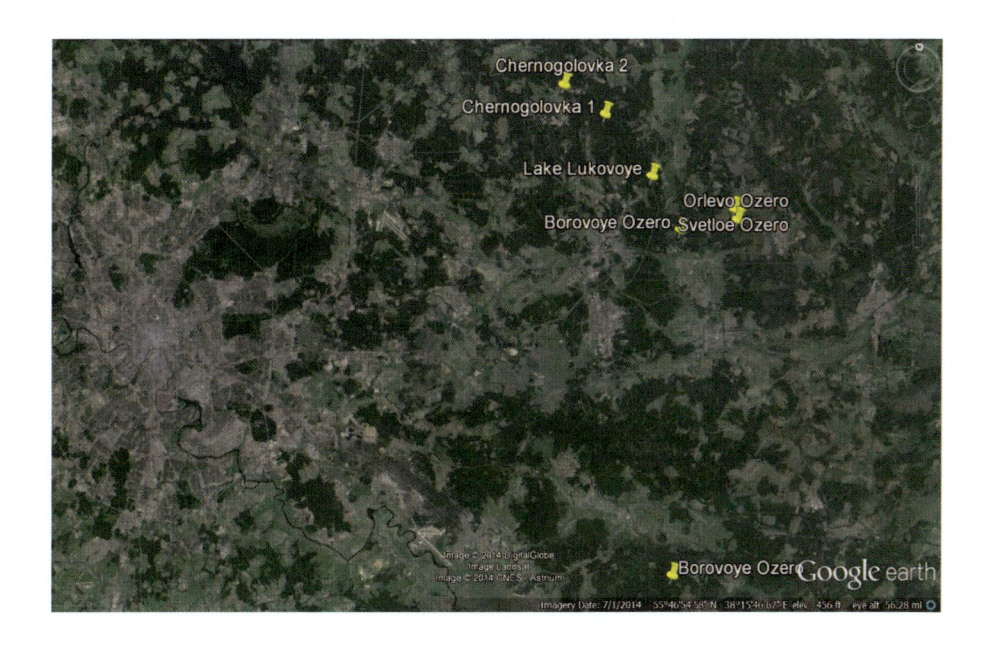

Svetloyar Crater is to the right and to the left is the larger cluster around Lake Lukovoye. Moscow is the urban area to the south-west.

Wait a minute? There was an even larger meteorite storm in this exact same area only three thousand years later around 8,000 BC? Like the Aral Crater with its two dates there was an even bigger meteorite storm around 8,000 BC? Once again is this too much for coincidence? Were these impacts actually from the massive meteor bombardment of 8,000 BC in the same area as the odds of two meteor swarms to hit the one pinpoint area on the face of the earth is more than astronomical? Again I ask, was the dating out? Chernolovka 2 is only seventy-four kilometres from the Svetetskiye impact. When you get to Russia in 8,000 BC you will see that this curve of impact craters is in the middle of later impacts. Are these a unique series of impacts onto one small area or the impacts onto one small area with the wrong dates on them? The majority of impacts appeared to be around ten thousand years ago.

The impacts from around 8,000 BC. The 11,000 BC impacts are the big cluster middle-left.

In 1964 the remains of American Red Indians along with their canoe and American made artifacts were discovered at Lake Ushkov in Siberia. These remains dated back to 11,000 BC. Were these an anthropological team sent to Siberia to seek out a theoretical route for their ancestors to have come across to enter the Americas, contrary to their legends, or were they merely travelers caught up in events unfolding at the time. Where did the American Indians come from anyway?

The Sunda Shelf or Sundaland lay where the southern reach of the South China Sea, the Gulf of Thailand and the Java Sea are all now. A land area equivalent to the Indian subcontinent vanished below the sea leaving only the Malay Archipelago, Indonesia, Indochina and Borneo protruding from it. Cores through the coral layers on the Sunda Shelf in the China Sea taken by Hanebuth and associates showed sudden changes in sea levels in this period indicating that the ice masses had melted quite rapidly and not gradually as originally theorized. There were two major jumps in radiocarbon levels around 11,000 BC and 15,000 BC. Fairbanks insists that this was a sea-level rise of fifty feet in a few weeks. This is the same as cores taken from Barbados and Tahiti that also indicate 11,000 BC. How devastating would a sudden fifty foot sea level rise be to our modern cities, many of which are built at sea level? Clube and Napier postulated that there had been a massive cometary or cosmic impact in this period. This would explain the sudden rise in sea levels and the evidence of massive tsunamis in this period.

Lake cores from southern Sweden show that rapid and sudden cooling occurred around 11,000 BC followed by very gradual drying. This cold snap lasted for one thousand years.

Under what are now the waters of Lake Assad behind the Tabqa Dam across the Euphrates River in Syria are the remains of one of the oldest organized communities of humans on the planet. This was Abu Hureyra which was a village of simple dwellings constructed around 11,500 BC and that had flourished in the high rainfall optimum period for over five hundred years. Abu Hureyra was composed of dwellings dug partially into the ground and then roofed with branches and patches of reeds supported by wooden posts. There had been generations of domestic occupation which contained thousands of seeds and other plant remains as well as fish bones and tiny beads. There were seeds from over one hundred and fifty types of edible plants and the community was originally ringed by open forests of oak, pistachio and other nuts that were within easy walking distance. Now the nearest forests are one hundred and fifty kilometers away. The remains of stonefruits and seeds were also found from hackberry, plum and medlar and white-flowered asphodel. There were long lines of pistachio trees on low wadi terraces near the village. This was an organized agricultural community in the Stone Age! The inhabitants also obtained wheat and two forms of rye that grew in boundaries between oak forests. The Abu Hureyrans also hunted gazelles and culled herds in large numbers.

By 11,000 BC the Abu Hureyrans had stopped gathering tree fruits and nuts from the forests as the forests retreated from the village. They became more dependent on feather grass and asphodel seeds which flourished in the open areas as the forest canopy retreated. By 10,600 BC even these grains vanished from the occupation layers and pistachio fruitlets became less common. The thousand year drought had set in and eventually the community disappeared as the area became dryer and dryer. Eventually by 10,000 BC the Abu Hureyrans had domesticated wild grasses such as rye, einkhorn which is a variety of coarse-grained wheat and lentils. These were not enough though to keep the community viable with no other food sources such as the nut trees that it used to have easy access to. Only a few generations after the domestication of the wild grasses the village was abandoned. Climate change had effected civilization though in this case man was not solely responsible but he did have a hand in it. As I said before a few times as cities grew the environment suffered and in many cases was desolated. Just look at Egypt and Mesopotamia now.

Cores through the coral layers in Tahiti taken by Bard and colleagues showed sudden changes in sea levels in this period indicating that the ice masses had melted quite rapidly and not gradually as originally theorized. There were two major jumps in radiocarbon levels around 11,000 BC and 15,000 BC. Fairbanks insists that this was a sealevel rise of fifty feet in a few weeks. This is the same as cores taken from Barbados and the Sunda Shelf in the China Sea that also indicate 11,000 BC.

From 1932 onwards projectiles similar to the Folsom ones were found on the border between Texas and New Mexico and west to Naco in Arizona. They are from ten to twelve centimetres long, 4.8 inches, and date back at least to 8,000 BC to 11,000 BC. They were able to penetrate the skull bone of a mammoth.

Also in 1958 near Lewisville stone tools and burned animal bones were found in association with hearths. These were radiocarbon dated as 38,000 years old. This appears to be a very old habitation area.

During the eleventh Millennium BC thousands of prehistoric animals and plants were mired, all at once, in the famous La Brea Tar Pits in Los Angeles. There were bison, horses, camels, sloths, mammoths, mastodons, and at least seven hundred saber-toothed tigers. One disarticulated human skeleton was found, completely enveloped in bitumen, mingled with the bones of an extinct species of vulture. The La Brea remains are broken, mashed, contorted, and mixed in an almost heterogeneous mass, which indicates a sudden and dreadful volcanic Cataclysm thirteen thousand years ago. Some scientists state that the extinctions in this area occurred as recently as 9,000 BC and occurred over twenty-five years. We can allow for this small geological dating variance when we compare it to the previous ones that we have seen. The human skull was typical of modern American Indians, not a prehistoric species of human.

Ancient man was active in quite a few places in this period. At Arlington Springs on Santa Rosa Island in Santa Barbara County in the Channel Islands of California the bones of a woman dating back thirteen thousand years were found. The bones were discovered in 1959. Does this indicate that people in this period were capable of travelling across water? Or were the Channel Islands connected to the mainland of California at this time? Some sources say yes and others say no. Geology is not an exact science.

In Louisville or Lewisville in Denton County in Texas from 1932 onwards projectiles similar to the Folsom ones were found on the border between Texas and New Mexico and west to Naco in Arizona. They are from ten to twelve centimetres long, 4.8 inches, and date back at least to 8,000 BC to 11,000 BC. They were able to penetrate the skull bone of a mammoth. In 1958 near Lewisville stone tools and burned animal bones were found in association with hearths. These were radiocarbon dated as 38,000 years old.

The Potomac Crater in Virginia in the United States is thirteen thousand years old and sixteen kilometres in diameter.

Remember that one hundred years is nothing geologically, possibly in reality impossible to actually determine or define so when you look at the impactoid events of 10,900 BC that follow then you can reasonably believe that they all occurred at the same time. We need to compress time periods to get rid of geological dating anomalies. It gets even more amazing when you squeeze them together. If it is good enough for the goelogists it is good enough for us.

How many impacts were there around 11,000 BC? We have the Aral Crater which is 450 kilometres across? There is the group of impacts near Moscow and the Potomac Crater in Virginia? Is this unusual activity? How many meteoric impacts leaving a crater more then half a kilometer in diameter can you recall in the past thousand years? But were these impacts actually part of a much larger meteor swarm impact around 8,000 BC? Can we really include the Moscow craters? Were all of the impacts actually around the same time? It is too much for coincidence for them to have occurred separately otherwise we had two massive meteor showers that hit the same places on earth three thousand years apart.

10,900 BC. The Younger Dryas.

Let the games begin!

The Younger Dryas lasted from 10,900 BC to 9,700 BC and was a sudden period of sharp decline in temperatures over much of the Northern Hemisphere. This weather period is thought to be caused by an influx of fresh cold water from North America into the Atlantic Ocean. In the Southern Hemisphere there was a warming. What caused the Younger Dryas?

In 10,900 BC the Younger Dryas occurred causing a major drop in temperatures and the return to the Ice Age that then lasted twelve hundred years. The period of the Younger Dryas was from 10,900 BC to 9,700 BC. During the Younger Dryas temperatures fell and the rains became scarce throughout southwest Asia, Europe and Africa. The Younger Dryas is named after a polar flower that was commonplace in this period and whose pollen is frequently found in waterlogged deposits of the time.

World ocean levels had risen 130 metres up until the start of the Younger Dryas but were still well below present sea-levels.

Around 10,900 BC according to several scientists an object one kilometre across grazed the Earth and released massive amounts of heat that caused massive wildfires across the United States and Canada. A layer of charcoal and glasslike beads have been found from this period. Fires apparently melted large portions of the Laurentide Ice Shelf causing masses of water to flood down the Mississippi River and into the Gulf of Mexico where the mass of cold fresh water altered the currents of the Atlantic Ocean and the Gulf Stream. Remember as you read this that there is great variability in our dating and many of our events may have actually occurred at the same time. This is why I have merged these periods together.

During the Younger Dryas massive lakes in Africa dried up.

The Luanda Crater in Angola is one kilometre in diameter and is twelve thousand nine hundred years old.

During the Younger Dryas the level of Lake Van in Armenia dropped by eight hundred feet in only a few centuries due to the arid climate. From pollen samples and the ratio of the stable isotopes of oxygen in the shells of zebra mussels it was worked out that the change from dry to moist at the end of the Younger Dryas only took ten to fifty years. During the shrinkage of Lake Van the water in it and the peripheral marshes was too salty for humans to drink.

Originally the Black Sea was the Euxene Lake and it was one hundred and fifty metres lower than the Sea of Marmara which was an extension of the Mediterranean Sea through the Dardenelles. Prior to the Younger Dryas massive amounts of meltwater flowed out of the Euxene Lake via the Dardenelles into the Mediterranean. This had happened for two thousand years then the Younger Dryas occurred and it stopped. Then the water started evaporating from the lake and the outflow channel became clogged with mud and debris and gradually formed an earthen berm or shallow natural dam. This turned the Euxene Lake bracken and shallow and so low that it was one hundred and fifty metres below the level of the Sea of Marmara and the Mediterranean. During the Younger Dryas precipitation in and around the Euxene Lake, later to become the Black Sea, was so low that it was reduced to the point that the loss of water by evaporation from its surface exceeded the water received from the rivers and rainfall causing the water level to drop until it had fallen below the Sakarya outlet wherein the outflow ceased and the Euxene Lake became totally landlocked. The Sakarya Channel now being no longer connected to either the Euxene Lake or the Mediterranean Sea slowly collected mud and debris brought to it by the seasonal rains and the flooding of its streams which built up slowly forming an earthen dam. Then as the lake slowly drew down it exposed an old shelf upon which a thick accumulation of the remains of marine organisms and rich sediments brought down by the many rivers was exposed. The retreating waters driven by winds and tides sloshed back and forth removing the silt from the uppermost layers of the sediments thus leaving only fragile shells of delta mollusks that were left broken and bleaching in the sun. The sunbaked cracks filled with sand and the seeds of wild wheats native to the area and they took root. There were new moist depressions and valleys through which new sinuous river valleys were cut to the shoreline. Detritus was carried by the rivers and deposited at the new lake edge of the Euxene Lake that built new deltas and natural deltas that teemed with fish and became a verdant refuge for man and beast.

Was this another Eden? Was this actually the original Eden? It was not in the best of locations as you shall see later.

The Marcador Paleolagoons in Bolivia were created twelve thousand nine hundred years ago around 10,900 BC. The largest crater is twenty-five kilometres long. Most are now elliptical lakes. This was a cluster of meteoric impacts. These impact craters are near Marcador in Jose Ballivian Province in Beni Department in Bolivia. Just think? What if we join the 11,000 BC events to the 10,900 BC events?

The Botswana Paleolagoons are meteoric craters that were formed twelve thousand nine hundred years ago. The largest are 3.5 kilometres wide. This was also a cluster of meteoric impacts. These paleolagoons are near Lehututu in Kgalagadi District in the Kalahari Desert in Botswana.

More scattered impacts from the same period!

Talking about meteorite storms? Now we hit a doozy.

Impactoids crashed down onto Brazil around 10,900 BC!

The Parcial Crater in Alagoas in Brazil is twelve thousand nine hundred years old and three hundred and fifty metres in diameter. The Parcial Crater is near Palmeira dos Indios in Alagoas in Brazil.

The Boca da Mata Crater also in Alagoas in Brazil is an impact crater twelve thousand nine hundred years old. It is seven hundred metres wide.

The Fiera de Santana Strewn Field in Bahia in Brazil is a collection of impact craters that are twelve thousand nine hundred years old. They are up to five kilometres in diameter.

The Campo Alegre de Lourdes Crater is twelve thousand nine hundred years old. There are numerous other impact craters around this crater. This crater is also in Bahia in Brazil.

Still in Bahia in Brazil and we have another impact crater. The Remanso North Crater Field is twelve thousand nine hundred years old. There are numerous impact craters around the largest crater which is 700 metres in diameter.

More Bahia impacts occurred. The Remanso South Crater Field is twelve thousand nine hundred years old. There are numerous impact craters around the largest crater which is 1.5 kilometres in diameter.

The Casa Nova Crater near Remanso in Bahia in Brazil is twelve thousand nine hundred years old and five hundred metres across.

And you thought that it was amazing when you met one crater?

The Vitoria da Conquista Craters or paleolagoons near Vitoria da Conquista in Bahia in Brazil are twelve thousand nine hundred years old. The craters are up to one kilometre in diameter.

The Itaparica Crater or paleolagoon near Xique Xique in Bahia in Brazil is twelve thousand nine hundred years old. The crater is six kilometres long and is elliptical.

The Espirito Santo Crater Field in Brazil is near Nova Venecia in Brazil in the State of Espirito Santo. There are numerous meteor craters here that are twelve thousand nine hundred years old around Pinheiros and Sao Mateus to the north and east of Nova Venecia.

The Urucuia Crater Field in Minas Gerais in Brazil is twelve thousand nine hundred years old. There are impact craters here that range up to five hundred metres in diamater.

The Jabatoa dos Guarapes impact crater or paleolagoon in Pernambuco in Brazil is twelve thousand nine hundred years old and three kilometres in diameter.

The Cajueiro Crater is three hundred and fifty metres wide and is near Paudalho in Pernambuco State in Brazil. The crater is twelve thousand nine hundred years old.

The Paudalho Crater is one hundred and fifty metres wide and is near Paudalho in Pernambuco State in Brazil. The crater is twelve thousand nine hundred years old.

The Cruz Crater is twelve thousand years old and three hundred metres in diameter. It is near Santa Cruz da Baixa Verde in Pernambuco in Brazil.

The Juazeiro Crater is now under the city of Juazeiro in Pernambuco in Brazil. The crater is twelve thousand nine hundred years old and three hundred and fifty metres in diameter.

The Palmeiro dos Indios Crater is one hundred yards in diameter and twelve thousand nine hundred years old. Nearby are two other craters, one of which is four hundred metres across whilst the other is three hundred metres wide. The Palmeiro dos Indios Crater is in Pernambuco in Brazil.

The Petrolinas Crater in Pernambuco in Brazil is five hundred metres in diameter and twelve thousand nine hundred years old.

How many of these things are there?

The Lunardo Lagoon near Serra Talharda in Paraiba in Brazil is an elliptical impact crater that is twelve thousand nine hundred years old. The crater is four hundred metres in diameter.

What was going on in Brazil?

How would these meteorite showers be impacting the earth?

The Lagoa Salgada or Salty Lagoon crater field near Puxinana is in Rio Grande do Norte in Nordeste in Brazil. The largest of the nine craters is three hundred and fifty metres in diameter and they are twelve thousand nine hundred years old. Stromatalites have been found here as well as shattered quartz.

The Capivara Crater is one hundred metres across and was created twelve thousand nine hundred years ago. The Capivara Crater is near Sao Raimundo Nonato in Piaui State in Brazil.

The Quari Crater near Sao Raimundo Nonato is seventy metres across and was created twelve thousand nine hundred years ago. Sao Raimundo Nonato is in Piaui in Brazil.

Were we having a swarm of meteoric impacts in Brazil in this time as well?

So far the score is three meteoric swarms with one in Bolivia, one in Botswana and one swarm in Brazil.

This is not counting the previous trio of falls in Russia roughly dated at 11,000 BC. These might have been at the same time as well.

Were there more to come in this period?

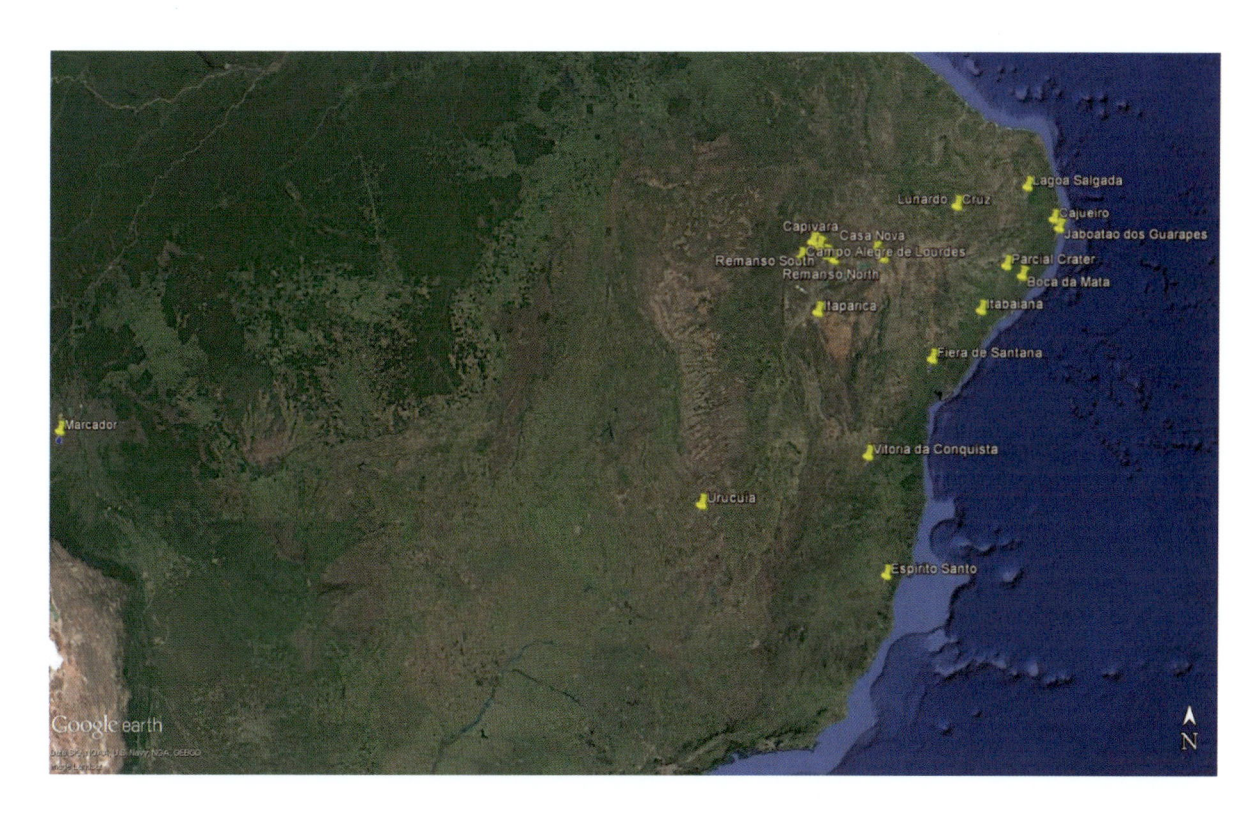

These are the Brazilian impacts of 10,900 BC.

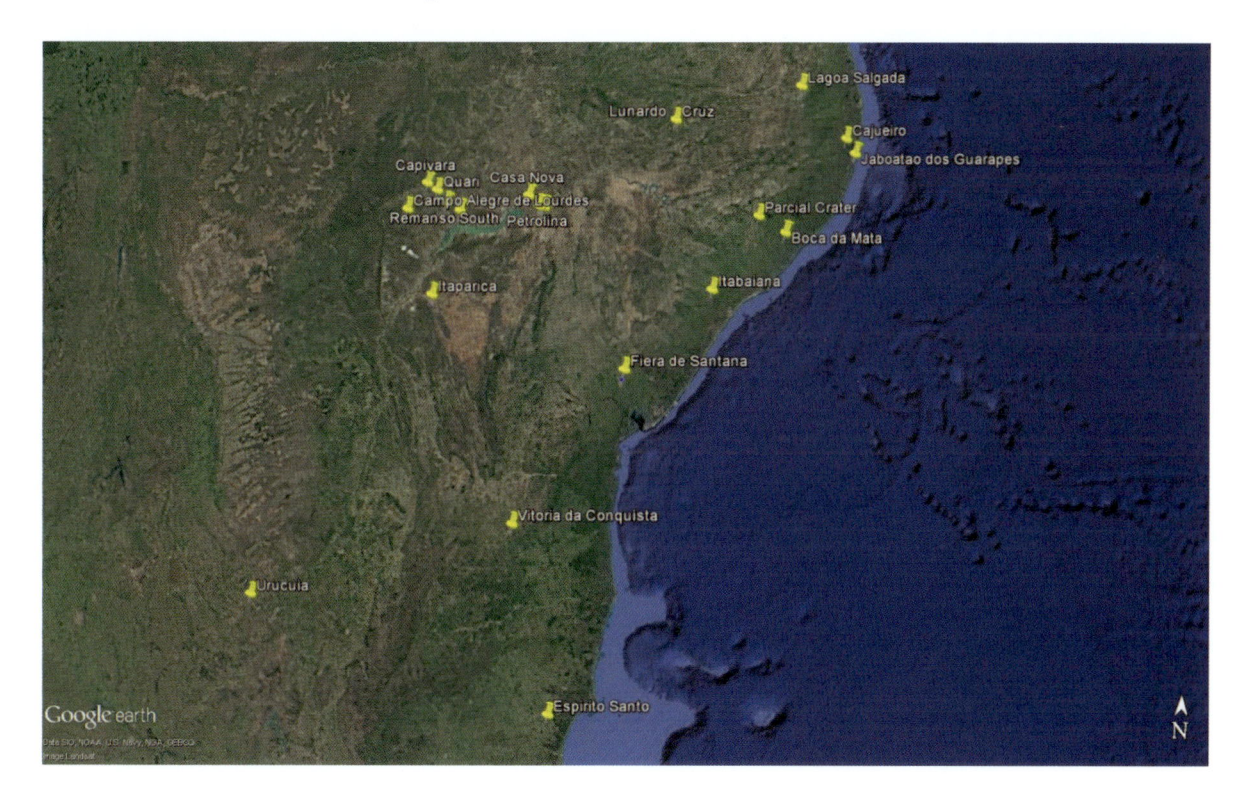

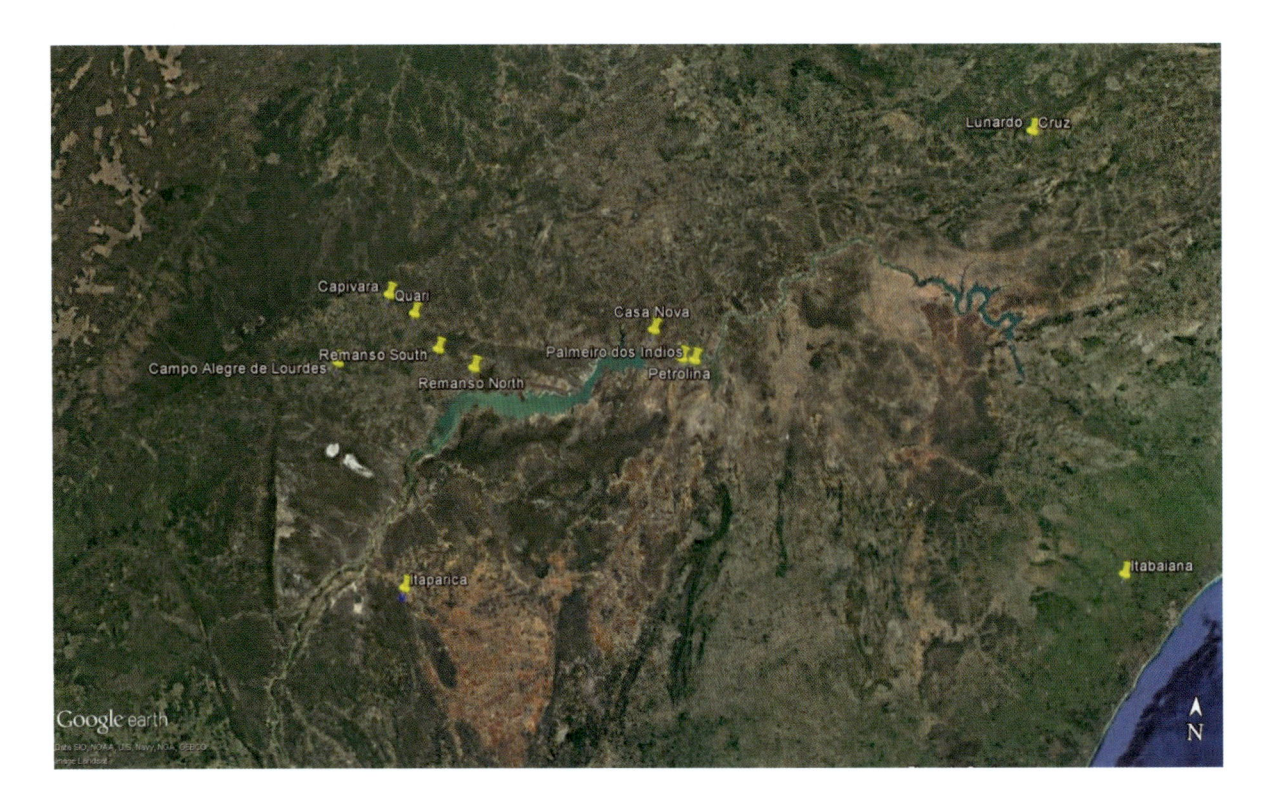

The previous three maps show the range of the Brazilian Impact Events in increasing detail.

The Itabaiana Strewn Field in Sergipe in Brazil is twelve thousand nine hundred years old and the largest crater is three hundred metres. The craters are elliptical indicating low angles of impact. These elliptical craters are called paleolagoons.

Botswana, Brazil and Bolivia are all on the same great curve of the Earth!

Now we finally leave Brazil and head for Canada. Should only be one here if that. And you still insist that nothing happened at the end of the last Ice Age?

Richard Firestone, Allen West and Simon Warwick-Smith state that the large raised rings discovered in Hudson Bay are in fact the remains of a mega-crater formed in 10,900 BC. The southeast shore is a uniform curve known as the Hudson Arc. This crater is two hundred miles wide or 320 kilometres by four hundred miles or 640 kilometres long and resembles a bulls-eye on the bottom of Hudson Bay. The edges of the ring are up to 250 feet high above ground level. When the impactor hit the area the ice sheet which was 10,000 feet thick suddenly had a huge hole in it. This caused meltwater to spurt out under the glaciers which caused massive flooding. This was one of the highest ice masses on the continent and yet in a very short time the ice sheet here disappeared. The Hudson Bay Ice Centre did not just melt away, it literally exploded. The explosion then shattered the rest of the ice sheet over Northern North America and ice masses ejected from the centre as well as shattering ice on the edges of the former ice sheet formed massive icebergs. Inland lakes were flooded as massive meltwater surges poured out from under the shattered and melting ice sheet. The ice dams of the huge glacial Lake Agassiz suddenly failed at the same time and masses of frigid meltwater surged towards the Atlantic Ocean possibly causing the Younger Dryas climate change. At the same time the circulation of the Atlantic Ocean shut down possibly by the massive influx of cold water from the meltwater surge from Lake Agassiz. At the end of the Younger Dryas the waters of the North Atlantic Ocean warmed by as much as seven degrees within a period of only fifty years. Once again this is one of these fifty year sudden temperature variations.

The Corossal Impact Crater near Sept-Iles in the Gulf of St Lawrence in Canada is a submerged impact crater that was created twelve thousand nine hundred years ago. The crater is four kilometres across.

But there are more Canadian impact sites.

At the Chobot farm near Buck Lake in Alberta in Canada blue-green algae mat has been found as well as nanodiamonds caused by a massive cosmic impact. Buck Lake is in Wetaskiwin County No 10 in Alberta.

Many other reports state that around 12,900 years ago a massive meteorite struck the Hudson Bay area and triggered a mini-Ice Age which we call the Younger Dryas. Other fragments of this same meteorite struck Alberta as well as Lake Hind, Manitoba. Other researchers put this impact at eleven to twelve thousand years ago, but in the wonderful world of geology that is nothing. Dark sediment found at Lake Hind indicates an impact event that caused global cooling and mass extinction in North America. The most recent dating of the event is fourteen thousand years ago and is indicated by the presence of nanodiamonds in sediment that has been exposed to extreme pressures and temperatures as would be caused by a massive explosion or impact caused by a cosmic body. These sediments were obtained from Murray Springs in Arizona, Bull Creek in Oklahoma, Gainey in Michigan aswell as Topper in South Carolina, Lake Hind in Manitoba and Chobot in Alberta. Lake Hind was a preglacial lake covering four thousand square kilometres in southwestern Manitoba that received meltwater from western Manitoba, Saskatchewan and North Dakota.

The Charity Shoal Crater is a shallow rimmed impact crater in Lake Ontario near Wolfe Island. The majority of the crater is in Canada and it is one thousand metres wide. The crater was created thirteen thousand years ago. Haven't we met this before? What is one hundred years anyway?

It was reported that a massive comet struck Northern Canada around 10,800 BC. This also apparently started off the Younger Dryas, a smaller Ice Age that lasted around 1,200 years. Massive forest fires were caused. This could well have been the same event recorded as occurring in 10,900 BC. In the cosmic and geologic scale of things one hundred years is not even an instant.

There was a massive meltwater event in Lake Agassiz in Manitoba in Canada at this same time.

Lake Agassiz was the largest of the many meltwater lakes in North America on the southern margins of the Laurentide ice sheet. Lake Agassiz in fact lapped onto 1,100 kilometres of the Laurentide Ice Sheet and covered parts of Manitoba, Ontario and Saskatchewan in Canada and Minnesota and North Dakota in the United States. A southwards bulge of the Laurentide ice shelf known as the Superior Lobe formed the eastern margin and it was this icy peninsula that blocked the lake waters from draining eastwards to the Atlantic Ocean via what is now the Saint Lawrence River Valley. Lake Aggasiz supported cold-loving mollusks that thrived in water temperatures of only five degrees centigrade. Lake Agassiz was so large a body of open water that its cold surface caused a strong southwards flow from the perennial high pressure areas over the ice to the north. This flow then blocked warmer winds and rainfall from the southwest and this combination of global warming at the time and minimal snow accumulation meant that the margins of the Laurentide ice sheet and the Superior Lobe began to retreat. Lake Agassiz entered a period of rapid growth being constantly filled by glacial meltwater. By 11,000 BC the lake extended so far eastwards that it almost flanked the southern edge. Normal geological theory then states that tiny rivulets of fresh water then crept across the Southern Lobe into what is now Lake Superior. These rivulets soon became deluges as a vast inundation of glacial meltwater crashed into the Saint Lawrence River Valley. Within weeks Lake Agassiz ceased to exist except in the form of remnants like Lake Winnipeg. This water then flowed into the Labrador Sea and this vast mass of much lighter freshwater floated on top of the much heavier saltwater. This unfortunately was on top of the Gulf Stream thus forming a lid that prevented warm water from cooling and sinking. The meltwater from Lake Agassiz suddenly turned off the Gulf Stream which influenced the climate of all of the countries around the Atlantic Ocean. Since the end of Heinrich One for the previous two thousand years saltwater downwelling in the southern Labrador Sea as well as off Iceland had propelled warm water from the Gulf Stream north and eastwards thus keeping Europe several degrees warmer than the equivalent latitudes elsewhere. This had now ceased. Within only a short period geologically temperatures dropped dramatically and the Scandinavian ice sheets started advancing again. At the same time a sea ice cap formed which prevented the Gulf Stream from starting up again. This was the period of the Younger Dryas with its return of Ice Age conditions.

Lake Agassiz, of which only a fragment remains in the present day flowed out to the North Atlantic as the Mississippi system was blocked off for up to one thousand years. Some scientists state that this instigated the Younger Dryas climatic change wherein the Earth reverted back to the Ice Age. Lake Agassiz was a very large glacial lake located in the middle of the northern part of North America with an area largers that that of all of the Great Lakes. The lake covered much of Manitoba, northwestern Ontario, northern Minnesota, eastern North Dakota as well as Saskatchewan with an area of 440,000 square kilometres. At various times Lake Agassiz drained out south through the Transverse Gap into the Minnesota River which was a tributary of the Mississippi River.

What would happen if this were to happen now? All Hell would break loose. And it did. You just have to compress the ages together and discover that Hell was here. And Hell has visited a few times and was about to make a major reappearance. We only mention Tunguska in common literature.

Above we see the Hudson Crater with Lake Hind and Chobot to its left. To the southeast are Corossal, Charity Shoal and Gainey in Michigan.

In Europe the Younger Dryas occurred causing a major drop in temperatures and the return to the Ice Age that lasted one thousand years. The period of the Younger Dryas was from 10,900 BC to 9,700 BC. This is the generally accepted date but as you would realize there are others.

Across Europe tree cover suddenly retreated and was replaced by smaller shrubs common to severe cold climatic conditions. This was accompanied by dramatic temperature fluctuations, severe climate swings and severe winter storms. World sea levels would have lowered in this period as water was converted into ice and snow and glaciers

With the Younger Dryas temperatures in Europe suddenly plunged. In Holland winter temperatures fell rapidly below minus twenty degrees centigrade and snow would fall between September and May with summer temperatures down to thirteen to fourteen degrees centigrade.

At the beginning of the Younger Dryas human and animal populations had to move from the sides of mountain ranges in the Negev Desert in Israel as the aridity of the Younger Dryas occurred leading to populations having to move to lower elevations as the game moved and the grasses shrank. The animals and humans in this case had to move back to the coast or to the Jordan Rift Valley whose floor is 1,200 feet below sea level.

The Natufians deserted the area where Jericho stood as well as many villages around it during the Younger Dryas. Jericho is in the West Bank.

Mexico has a verifiable impact crater from this period as well. The Manuel Benavides Crater in Chihuahua in Mexico is forty-five kilometres in diameter and twelve thousand nine hundred years old. Remember what I wrote about major impact craters in this period?

The vast plains of Russia during the Younger Dryas reverted to steppe desert due to the extreme dryness.

The Novosibirsk Crater Field in Siberia is twelve thousand nine hundred years old. The largest crater is one kilometre in diameter. These aren't just single impacts but fields of impacts. Were we sailing through a meteorite swarm or several meteoric swarms? How many swarms were there?

The vast plains of The Ukraine during the Younger Dryas reverted to steppe desert due to the extreme dryness.

Massive lakes dried up in Anatolia in Turkey during the Younger Dryas.

Lake Aksehir on the Konya Plateau in the west of Anatolia in Turkey was five times its present size as recently as thirteen thousand years ago but during the Younger Dryas it had shrunk down to a pond surrounded by cobbles and gravel that were high and dry.

Between eleven thousand and nine thousand BC, 11,000 BC to 9,000 BC, there were numerous upheavals in the northern regions of Siberia and Alaska around the edge of the Arctic Circle. Uncountable numbers of large animals have been found; many carcasses still intact as well as astonishing quantities of perfectly preserved mammoth tusks. Hundreds of thousands of individual creatures must have frozen immediately after death otherwise the meat and ivory would have spoiled. The mammoth meat appears so fresh that it is reported that it has been offered in restaurants in Fairbanks, Alaska, and has been used to feed sled dogs. Interspersed within the muck depths and sometimes through the very piles of bones and tusks themselves are layers of volcanic ash. This indicates volcanic eruptions of tremendous proportions at the time of the sudden deaths of the animals.

In this same period the glacial ice sheet suddenly retreated from Lake Superior and went back into Canada thus ending the last Ice Age. The line where the ice sheet had stopped originally is called the Mason-Quimby line and stretched from Lake Michigan to Lake Huron in Michigan. Below this line Clovis era spear points and remains of megafauna were found and not above.

In Murray Springs in the San Pedro Riparian National Conservation Area in Arizona nanodiamonds found in sediment indicate meteorite impacts around 10,900 BC. Other nanodiamonds were found in Bull Creek in Oklahoma, Gainey in Michigan as well as Topper in South Carolina, Lake Hind in Manitoba and Chobot in Alberta. Was this a North American swarm of meteorites?

Nanondiamonds were found in Topper in Allendale County in South Carolina. There appeared to be an industrial scale toolmaking site that was in existence 16,000 to 20,000 years ago in this area as well.

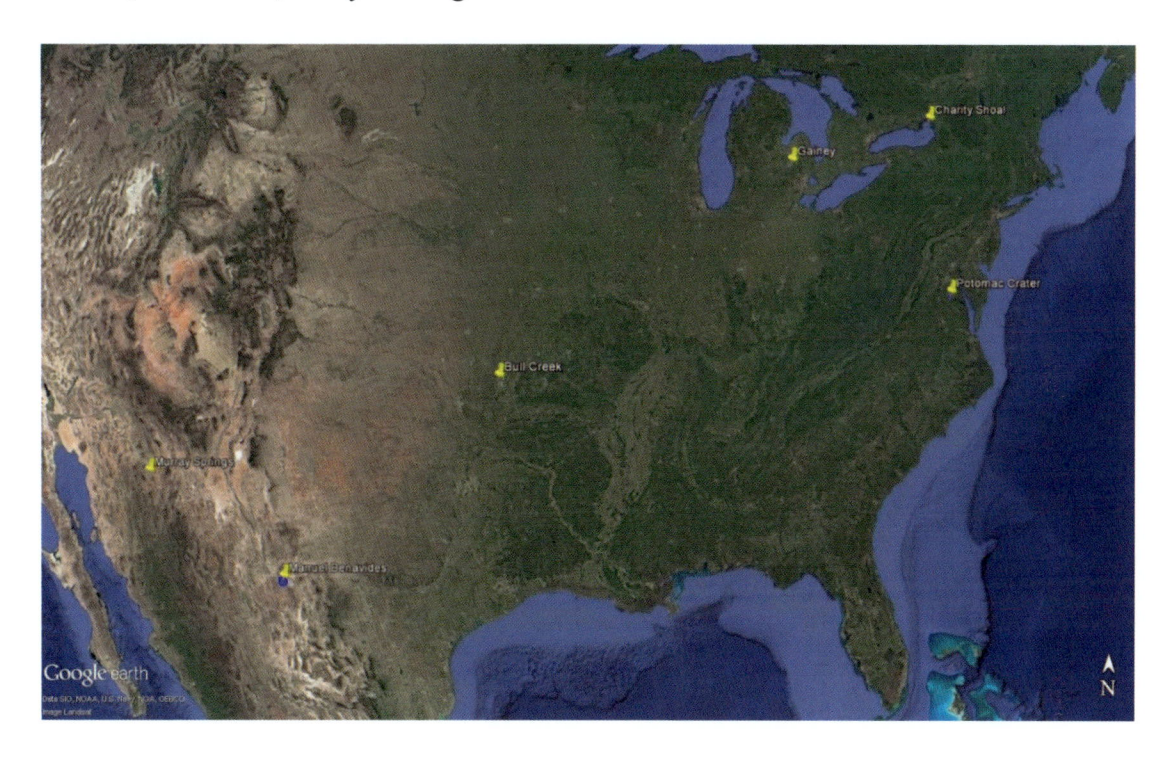

Sites of impact nanodiamonds at Murray Springs, Manuel Benavides, Bull Creek and craters at Gainey, Charity Shoal and the Potomac Crater.

The combined craters and nanodiamond fields of North America around 10,900 BC.

In the Kaoma District of Western Province in Zambia are numerous paleolagoons or elliptical impact craters that are around twelve thousand nine hundred years old. The largest are up to five kilometres across. The Kaoma Craters are near Mongu.

Always allow for dating variance though. Most datings allow several hundred years variance.

There is a persistent black band of charcoal-rich material found in England dating back to 10,770 BC. This indicates massive conflagrations across the country in this period.

The sudden melting of glaciers around 10,700 BC in southwestern Ontario in Canada in North America caused a massive deluge that released two thousand square miles of freshwater into the Atlantic Ocean. This apparently influenced the Atlantic Ocean thermohaline circulation and sent temperatures plummeting.

We could even allow for five hundred year time increments as being the one occasion especially if we have not allowed for radiation spikes in our dating systems.

Temperatures over Greenland around 10,700 BC suddenly dropped eighteen degrees Fahrenheit on average. How fast was this though?

Around 10,550 BC a massive conflagration occurred in the Nile Valley of Egypt. There is a layer of burnt material on top of the uppermost layer of the Sahaba Silt. There seemed to be a massive fire for two hundred kilometres along both sides of the Nile River. This layer of fire damage stretched from

Esna northwards to beyond Qena. The geologists have allowed for two hundred and thirty years either side of this date which would include the same type of burnt layer in England allegedly in 10,770 BC. What would cause simultaneous massive firestorms at the same time in two different areas not exactly renowned for bush fires? Remember that even two hundred years in geology is still the same time.

Around 10,500 BC the earth's climate suddenly rose by over twenty degrees Fahrenheit and ended an Ice Age. Ice cores from Greenland show that there was a rise of 59 degrees Fahrenheit in what is now the North Polar Region for around fifty years. Another fifty year temperature increase event! Or is it the same one that we read of earlier?

Six miles from Alice Springs at Ndahla Gorge in the middle of one of the most inhospitable areas on Earth there are drawings of godlike figures with what appear to be gigantic antennae on their heads. Robert Edwards also discovered the faces of Gods wearing protective goggles engraved on the rock. Lines that cross each other or run parallel to each other only to end abruptly are engraved on a rock four feet seven inches long and three feet wide. They resemble the network of lines on the plain of Nazca in Peru, only in miniature. Alice Springs is in the Northern Territory in Australia and the carvings date back to 10,500 BC.

On a rocky cliff face West of Alice Springs Michael Terry discovered a carving of the extinct *Nototherium Mitchelli* in 1962. This Rhinoceros type creature had disappeared 12,500 years ago, 10,500 BC. In the same place Mr Terry also found six carvings of what looked like Ram's heads in the Assyrian style. A human being about two metres tall was also found. He had full legs and thighs and held a mitre resembling those worn by the Pharaohs. However, the figure is in a horizontal position and it is standing as if walking down a wall. The ram was unknown in Australia until the arrival of the Europeans in the late eighteenth century. Signs of erosion on the carvings speak of great age. He could also be a Birdman, similar to those of Easter Island. Alternatively, is it Thoth, the Egyptian Ibis-headed god, again?

Around 10,500 BC the Beringian land bridge had been severed and the Pacific and Arctic Oceans met for the first time in many millennia. What had caused the sudden increase in sea water levels to do this? The Beringian landbridge was between Alaska and Siberia. This is usually the result of glacial melting. The land had not dropped. The sea had risen. Yet this is near the middle of the Younger Dryas when sea levels were apparently falling due to increases in glaciation. Someone has their dates wrong or there were geological reasons for this as it could not be from massive sea level rises in this period.

By 10,500 BC we have already had a massive meteoric fall in Canada as well as the flooding release of Lake Agassiz as well as the sudden glacial melt event as well in Canada and North America. Had these influenced the rise in world sea levels? Or were there other factors? Were they the mass impactoid events around 10,900 BC? Is the 10,900 BC date wrong? Dating is very erratic in this period of the Cataclysm.

Between 10,500 BC and 9,500 BC there were extremely high Nile Floods that continually engulfed the Nile Valley. Where is this water coming from during the great drought?

Shanidar is a huge cave situated high above the Greater Zab River about 520 kilometres northeast of Abu Hureyra in the mountains of Kurdistan. It has a long history. Found at Shanidar which appeared to have been a summer encampment there have been found thousands of broken animal bones from immature wild sheep dating to 10,500 BC. This seems to show that the inhabitants practiced careful selection of wildstock. Ralph Solecki also discovered a slim almond shaped piece of copper with two equally spaced perforations at its end so that it could be worn as a pendant around the neck. The stratum it was found in was around 9,500 BC. You may have noticed that some articles appear to be repeating themselves but this is merely to emphasize that some sites have various dated discoveries as well as long histories.

There was an interesting discovery in Mud Lake, also called Friendship Lake, in Kenosha County in Wisconsin. In January 1936 a Works Progress Administration crew was digging a drainage ditch and unearthed most of the foreleg of a juvenile mammoth. In 1991 an amateur archeologist noted cut marks on the bone similar to other mammoth bones found at Fenske and Schaefer sites nearby. In 1992 and 1993 archeologists excavated Shaefer site and discovered bones with cut marks on them as well as stone tools underneath a pelvis bone. Radiocarbon dates on the bones and on plant material at the site of the dig ranged from 12,500 to 12,300 years ago.

In 10,376 BC the Gothenburg Magnetic Flip occurred. Recent studies in the field of Paleomagnetism, fossil magnetism, indicate that the Earth's magnetic polarity has reversed itself more than 170 times in the last eighty million years. To help explain what this means think of the Earth suddenly lurching or tilting. The Jury is still out on how fast the planet does this but is unanimous on the fact that it does. The Earth's magnetic fields reversed in what is known as the Gothenburg Magnetic Flip in 10,376 BC, nearly 12,400 years ago. When there is a magnetic field reversal or flip the Earth's magnetic field drops to null or zero which minimizes the shielding effect of the radiation against cosmic radiation which would allow increased cosmic particle penetration or radiation of the Earth's atmosphere. This cosmic radiation would also affect our radiation decay readings. Are our dates correct at all?

According to studies of fossil magnetism around 10,400 BC, there was a 180-degree reversal of the Earths Magnetic Poles. I think we can agree that this was the Gothenburg magnetic Flip.

The Geologist S. K. Runcorn of Cambridge University made the point that there seems no doubt that the Earth's magnetic field is tied up in some way to the rotation of the planet and the conclusion is that the Earth's axis of rotation has changed also so the planet has rolled about thus changing the location of the geographical poles.

According to reports published in "Nature" and "New Scientist" the last geomagnetic reversal was 12,400 years ago, 10,400 BC, during the eleventh Millennium BC. This is the same date as the Gothenburg Magnetic Flip. Some Scientists predict that the next reversal of the Magnetic Poles will occur in 2,030 AD.

In 1958 during the International Geophysical Year research indicated that the Earth was in fact pear shaped. In the Atlantic area the bulge lies a great deal further south. This could possibly have been the equator before the tilting of the Earth's axis 12,000 years ago, around 10,000 BC. Then the North Pole was located somewhere near the longitude and latitude of Iceland or the south end of Greenland making the climate along the line London-Berlin-Warsaw the same as present day Spitzbergen. The Bering straits were as far from this pole as the Dardenelles between Europe and Asia in Turkey which link the Black Sea to the Aegean.

You will find periods repeating themselves almost ad nauseum.

Legrand and De Angelis sampled ice cores from GRIP from Greenland that indicated that there was a massive surge in ammonium levels in the ice cores. They had begun to rise in 14,000 BC but in 11,000 BC they escalated from an average over 400,000 years of a few parts per billion to sixty-three parts per billion indicating a huge increase. The ammonium levels are indicative of massive fires consuming the Earth in this period. The actual date of the spike is 10,340 BC. There had been a similar but smaller spike in 39,000 BC. Another theory is that the ammonia arrived via comets and cometary dust which also contain ammonium. We certainly had the comets for this to occur during this period.

Legrand and De Angelis state that also in the period of 10,340 BC there was a massive surge in nitrates into our atmosphere as found in Greenland ice cores. These nitrates only appear after massive firestorms.

Richard Firestone, Allen West and Simon Warwick-Smith state that there was an increase of bombardment of the Earth around 10,340 BC by supernova particles and debris that had first hit the Earth in 39,000 BC. In these bombardment clouds were masses of particles with the density of styrofoam but many miles across as they accreted due to gravitation in their journey from the exploding supernova. They had intense velocity though they had little mass and there were tens of thousands of them. Some of these hit the moon before hitting the Earth and would have resembled explosions on the moon that were generating intense light. The inhabitants of the Earth would have headed for shelter this time as there would have been some warning. The gigantic cosmic balls would have crashed into every planet as well as the Sun where they would have exploded and created massive solar flares many of which would have hit the Earth. The horizon of the Earth would have resembled a cascade of comets.

Brilliantly illuminated the dust and ice balls would shine like the sun and though many would explode in the atmosphere many would collide with Earth. The largest fragment is believed to have crashed through the Glaciers over Hudson Bay in Canada and created a gigantic hole whilst others hit the Great Lakes as well as Siberia and northern Europe. Massive earthquakes shook the Northern Hemisphere from the impacts. This was followed by a blast of superheated air travelling at over one thousand miles per hour that tore trees from the ground and flash scorching animals, people and the earth. The only survivors were those who were underground or in other sheltered places. Micrometeoric particles peppered the Earth. The collision with the ice sheet caused the sudden meltwater event under the glaciers that created the drumlin fields. The massive cold climate grasslands burnt depriving surviving animals of food. The thunderous impacts had shocked the magnetic fields of our planet so that they wavered causing the magnetic poles to wander erratically across the planet with the North Magnetic Pole almost hitting the equator before it restored its equilibrium. Ice from the Hudson Bay impact was hurled high into the air and came crashing down forming the Carolina Bays. Other flying ice lumps hit Nebraska, Kansas, Texas, Oklahoma, Arizona and New Mexico with some ice masses hitting California and Mexico. Large parts of Europe and Asia were barraged by the ice masses. The disintegrating glaciers then moved along on the exploding meltwater causing havoc in front of them as well as massive flooding and tsunamis. Then it started raining incessantly as water vapour gathered around dust particles in the atmosphere. Eventually the lakes and bogs caused by the rain spawned the toxic algae that covered the extinct megafauna as well as the Clovis sites. Clovis sites are named after North American archaeological sites where a particular type of spearpoint was created or used that were all flaked in a particular way. As well ice dams collapsed due to the meltwater from the glaciers and glacial lakes overflowed with massive tsunamis on land. The sudden chill that accompanied all of this created the Younger Dryas that lasted over one thousand years in a return to the Ice Ages.

    Richard Firestone, Allen West and Simon Warwick-Smith have found that by mapping meteorite falls in North America there are streaks of meteorite falls that radiate out from the Great Lakes region and that much of the meteoric debris lies on top of geologic formations dating from the end of the last Ice Age. This indicates that the meteorite fragments or ejecta that stretch out up to 2,400 miles from the impact site to California as one example were caused by a massive impact in the Great Lakes region. In Canada they stretch for one thousand miles and another series stretched across Texas and almost to the Mexican border. All of North America seems to have been covered with high impact hot meteoric debris.

    In North America, Ireland, Lithuania, Estonia, Poland, Finland and Denmark there are geologic anomalies called drumlins which were formed in this period and at no other time. Some scientists theorize that drumlins were formed by meltwater flooding of the massive glaciers at the end of the last Ice Age. Drumlins are boat shaped mounds of sand, rock and gravel and are from a few hundred feet long to five miles in length and up to one mile wide with most being less than one hundred feet high. One theory states that drumlins were formed when water pressure under a heavy ice sheet or glacier periodically bursts out in a huge flood carving 360 degree channels under the glacier that are then filled with debris being carried along. Drumlins were only produced in this period and the explosions of water for there are massive fields of these drumlins could have been caused by the crashing down of a cosmic body onto the centre of a glacial sheet forcing the water to be forced outwards. The high velocity surge of subglacial muddy water carved depressions in the bottom of the ice sheet in the shape of drumlins and when the flow stopped these cavities filled with debris such as gravel and sand amongst other things. After the glacier retreated or melted the drumlins remained. They have not been found in association with earlier Ice Ages either so are not normal Ice Age phenomena, only phenomena from the last Ice Age, occurring at the end of it. Drumlins are not being formed today. The glaciers melted away above the drumlins for if the Glacier had moved over them it would have acted like a gigantic bulldozer and taken them with it before flattening them.

 Around 10,178 BC the Celestial Pole was inclined at an angle of thirty degrees from its present position which indicates that the terrestrial axis was oriented differently from today. This indicates a sudden physical tilt of the Earth's axis. Was this caused by the celestial impact around 10,340 BC that possibly resulted in the Gothenburg Magnetic Flip?

 The tally of meteorite impacts in the Eleventh Millennium. We had the undiscovered Labrador Crater which spread impact detritus over a huge area of North America, the Aral Impact Crater in Kazakhstan, the Svetloyar Ozero in East Nizhny Novgorod, the Borovoye Ozero near Moscow, Chernogolvka 1 and Chernogolovka 2 near Moscow, Lake Lukovoye also near Moscow and Orlevo also in the Moscow region, the Potomac Crater in Virginia, the Luanda Crater in Angola, the Marcador Paleolagoons in Bolivia, the Botswana Paleolagoons, the enormous Brazilian paleolagoons, Hudson Bay, the Corossal Impact Crater in Canada, the Charity Shoal Crater in Canada, the Manuel Benavides Crater in Mexico, the Novosibirsk Crater Field in Siberia and the Kaoma Craters in Zambia.

 Why the sudden explosions of meteorite impacts in the Eleventh Millenium? Or do we need to merge the four milleniums from the eleventh to the eighth as the falls appear to repeat at the same geographical places again? Is this too much for statistical reality?

10th Millenium BC.

Remember that when you see 10,000 BC, 9,000 BC and 8,000 BC that the dating is very approximate and can mean any time in each thousand year period. Many scientists used to state when dating events that they occurred around ten thousand or around nine thousand BC. This could mean that the events of three thousand years might have occurred around the same time.

The Polish scientist Professor Ludwig Zeidler of Poland believed that the destruction of Atlantis was due to a large cosmic body which struck our Earth about 12,000 years ago causing the earth to shift 23 degrees on its axis. This sudden change froze mammoths solid that were feeding in Siberia as their then temperate mid-Summer savannahs suddenly became Arctic ice deserts as the Arctic weather enveloped them with rapidly freezing temperatures. This area was not tundra then, it was temperate savannah.

The North Pole moved from Hudson Bay in Canada to its present location around 12,000 years ago.

There were elephants roaming the plains of North America and South America prior to twelve thousand years ago. The turmoil that caused the extinction of so many species of animals both in the New and Old Worlds according to Charles Darwin must have shaken the whole framework of the Earth. In the New World more then seventy genera of large mammals became extinct between 15,000 BC and 7,000 BC. This included all North American members of seven families and one complete order, *Proboscidea* or elephants. These staggering losses involving the violent obliteration of more than forty million animals were not spread out evenly over the entire period but were in a small period of two thousand years between 11,000 BC and 9,000 BC. In perspective in the previous three hundred thousand years only about twenty genera had disappeared. This pattern was repeated across Europe and Asia. Even far off Australia was not exempt losing around nineteen genera of large vertebrates, not all mammals, in this relatively short period of time.

Blue and green crystal which can only be produced by nuclear explosion or an extreme source of heat vitrifying the Earth or sand under the explosion site was last formed all over the world during the period of 10,000 BC to 9,000 BC. We only duplicated this type of crystal with our own nuclear explosions in the late 1940s.

In the International Geophysical Year American scientists fished up from the bed of the Antarctic Ocean specimens of muddy sediment, which showed that in comparatively recent times Antarctic rivers had borne alluvial products of an ice-free terrain. This was the case ten to twelve thousand years ago.

The Russian hydrologist M. Ermolaev showed that the present Arctic water system was established around 12,000 years ago which was the end of the glacial epoch in Europe and North America.

Some of the islands of the Arctic Ocean were never covered by ice in the last Ice Age. On Baffin Island nine hundred miles from the North Pole alder and birch remains found in peat suggest a much warmer climate less than thirty thousand years ago. These conditions prevailed until seventeen thousand years ago. During the Wisconsin Glaciation or Wisconsin Ice Age there was a temperate climate refuge in the middle of the Arctic Ocean for the flora and fauna that could not exist in Canada and the United States.

Russian Scientists concluded that the Arctic Ocean was warm during most of the last Ice Age and the period 32,000 to 18,000 years ago as being particularly warm.

On Kotelnyy Island in the Arctic dead trees stand where they grew thousands of years ago. Their tops are twisted and smashed or they lie in confused mounds of timber up to one hundred and eighty feet deep as if a gigantic bulldozer had pushed them forwards. Or was this a mud and water megatsunami?

Above the Arctic Circle are the Queen Elizabeth Islands which are desolate and treeless and freezing cold even in summer. Enormous quantities of drift are found here that are jumbled masses of stone and plant matter that comprise leaves, pinecones and acorns and the piles of drift are up to three hundred feet above sea level in jumbled masses up to forty feet deep. This area had a warmer climate in this period.

A reindeer's skull has been found in Lake Sevan in Armenia which is a mystery as they are creatures of the plains and not of the Caucasus Mountains. The skull is 12,000 years old. Had the plains suddenly been raised as our world geology changed suddenly? Indications are that the Caucasus Mountain range rose up during the last Cataclysm being a plain beforehand. Thousands of reindeer perished here en masse as the area rose up to its present height of 5,000 feet where reindeer cannot survive today.

There are two schools of Geology which are the Gradualist School and the Catastrophist School. Which school is right or is it a mixture of both?

About twelve thousand years ago, 10,000 BC, the rate of sedimentation on the sea bottom of the Atlantic Ocean was considerably reduced while at the same time an abrupt change of climate affected the whole Planet.

The American Geophysicist Piggot collected soil samples from the Atlantic Ocean that consisted of stratigraphic cores of nine feet ten inches in length and frequently including two zones very rich in volcanic ash. The ash must have originated in enormous volcanic eruptions either in the West Indies or in the central ridge of the Atlantic. The topmost of the two volcanic strata is found above the topmost glacial stratum indicating that the volcanic eruption or catastrophe occurred in postglacial times or at least at the end of glacial times.

In 1956 Dr P. W. Kolbe of the Riks Museum in Stockholm, Sweden, found proof of a sinking of the Mid-Atlantic Ridge. A core sample had been taken at a depth of 12,000 feet. The shells of diatoms were found in it. Diatoms are miniscule marine animals and the remains of the variety found were of a freshwater type that could have only been deposited there when the area was a lake. The present sea-bottom had to have been above sea level at one time. This would have been 10,000 to 12,000 years ago according to Dr Malaise, also of the Riks Museum.

Dr Kolbe reached depths of 3,700 metres and found flint deposits, which he brought to the surface. Kolbe, joined by R. Malaise, concluded that these vegetation residues had been located in sweet water, not saltwater, about 10,000 to 12,000 years ago. Something large had crashed to the bottom of the Atlantic Ocean.

Australia, New Guinea and Tasmania were all one continent up until twelve to ten thousand years ago. A thin line separated it from Sundaland on the opposite shore of the Wallis Line which is home to excessively powerful cross currents. Nothing can just float across the Wallis Line. It has to be intentionally manouvred across as this is the demarcation point between the land of *mammals* in Southeast Asia and the world of the *marsupials* and *monotremes* in Australia and Niugini.

At Alcoota, 150 kilometres northwest from Alice Springs, in Australia, the fossil remains of giant geese weighing up to five hundred pounds were discovered. There were three species of giant goose at the site. Two smaller types weighing between 150 and 200 kilograms and the larger *dromornis stirtoni* at over 500 kilograms or half a ton! These giant and flightless birds lived from fifteen million years ago to just 30,000 years ago when the local environment changed from forests and grasslands possessing a plentiful water supply to the desert of today. Some state that this desertification did not occur until 10,000 BC. This was the same period when the inland seas of Australia suddenly dried up.

The Azores Islands are the tops of very tall mountains, rising over twenty thousand feet above the Atlantic seabed plains.

Where the Gulf Stream now flows there is a submerged island shelf around the Azores covering an area of approximately four hundred thousand kilometres that became submerged around 10,000 BC. Before the rise of world sea levels this could very well have blocked the passage of the Gulf Stream to the west of Europe and the British Isles which would have caused glaciation to occur in Western Europe and the British Isles. Then the landmass around the Azores sank and the Gulf Stream commenced heating up the eastern Atlantic Ocean. This landmass suddenly sank to a depth of three kilometres!

On top of one of the submarine peaks in the Atlantic Ocean a number of strange limestone discs have been found with a diameter of about fifteen centimetres and a thickness of about four centimetres. They were smooth enough in parts but elsewhere they were marked with rough indentations as if they might be plates. Radioactive carbon dating revealed that they were above water 12,000 years ago, 10,000 BC. But as we have shown already our radio carbon dating might not be that accurate after all. A submarine probe by the Geological Society of America in 1949 discovered the discs. There were about one ton of them on the bed of the Atlantic just south of the Azores. There were peculiar indentations in the centre of them. They were not natural formations and could not be identified. The state of lithification of the limestone indicated that it had been lithified under aerial conditions and that the seamount may have been an island within the last 12,000 years. Were these Neolithic ritual items? If so what was early man doing in the middle of the Atlantic Ocean?

We forget that ancient man used the seas, oceans and rivers as highways as overland was too dangerous and took too long. You could also handle larger cargoes using water traffic. The ancient seas were no barrier to ancient man.

In the Bahamas there are several unusual blue holes, not caves, but underwater circular holes half a mile in diameter going straight down to depths of one thousand feet although the sea bottom there is only several fathoms deep. It appears as if large limestone cave systems have collapsed suddenly. Are these holes related to those in the Carolinas and also off Puerto Rico that are evidence of meteorite bombardment around the same time around 10,000 BC?

In addition from this period we have a gigantic meteoric or asteroid hole in the Lesser Antilles as well as two meteoric holes near Puerto Rico. Are these holes all related? Were they all created by meteorites crashing through shallow limestone roofs of massive cave systems?

Interesting ruins from twelve thousand years ago have apparently been discovered off Andros Island in the Bahamas. In an area between 23 degrees 50 minutes and 23 degrees 30 minutes north and 80 degrees 30 to 79 degrees 40 minutes west extensive stone pavements have been found and photographed at a depth of 25 feet. There were also distinct walls with vestiges of pavement running along the top. The main wall runs for a quarter of a mile out to sea where it suddenly disappears into 2,500 feet of water. Part of the wall bifurcates near the shore and continues underwater partially along the coastline of what was once a much larger island now under the ocean. At another point on this submarine plateau divers followed a passageway under submerged rocks and discovered a sunken quarry complete with shaped rocks still inside it. The entire underwater plateau where these remains have been found is about sixty miles on each side of a lopsided triangle between the Strait of Florida and the Santaren Channel and Nicholas Channel. It breaks the water only at its edges and has many springs the same as in the Azores. Just within this area there are several unusual blue holes, not caves, but underwater circular holes half a mile in diameter going straight down to depths of one thousand feet although the sea bottom there is only several fathoms deep. It appears as if large limestone cave systems have collapsed suddenly but each time leaving these circular holes. You would think that when a cave system collapses it would follow the shape of the cave system and resemble a canyon. But these areas are circular, both on land and in the sea. Are these holes related to those in the Carolinas and also off Puerto Rico that are evidence of meteorite bombardment around 10,000 BC?

The deep holes in the ocean can be compared to the sacrificial wells used by the Mayas where they used to throw gold, jade and maidens as sacrifices to the Gods. Had ancestors of the Mayas remembered what caused the circular holes in the first place? Were these the landing places of the gods which were probably meteorites originally?

Along the coastline of Andros Island the French undersea explorer Jacques Cousteau found a huge stalactite and stalagmite cave that could only have been formed by drops of water falling over long periods of time in free air and not under water. The sediment in the cave was over twelve thousand years old. The huge Andros Grotto was 165 feet under water and had stalactites and stalagmites. On the cliff-like eastern side of Andros Island divers have felt their way down the base of the island and have found underwater caverns and grottoes studded with stalactites and stalagmites that can be only formed by slow trickle and evaporation of mineral bearing water, which is impossible under the sea.

The whole Bahama Banks is a landmass that was dry land as recently as eleven thousand years ago, 9,000 BC. If one were to reduce the sea level by fifty feet then there would be one large island indented by the Tongue of the Ocean up to the side of where Andros would be.

Coral sequences in Barbados showed the sudden onset of abrupt refrigeration with the cold dry climate of the Younger Dryas producing a drastic slowing of the rise of global sea levels.

Jacques Cousteau discovered the Blue Hole, a deep abyss near the coast of Belize in Central America. Here he discovered a labyrinth of stalactite and stalagmite caves at an angle that is impossible for them to form naturally. A strong earthquake must have tilted these caves and their calcite deposits. Analysis of the stalactites showed their age to be twelve thousand years. Belize was formerly British Honduras and the hole is one thousand feet wide in a labyrinth of underwater caves that could only have been formed above ground. In the sea off Belize there are several visible patches in the sea, as well as the Blue Hole, that are always a different colour to that of the surrounding sea, usually a turquoize blue. The patches are above holes in the seabed that are entrances to a limestone cave system that was above sea-level prior to 10,000 BC. What caused the sea to cover the cave system? Did the caves drop whilst the sea levels rose? Average sea level rises in this period is determined to be three hundred feet.

There is clear evidence that the Caribbean Sea and the Gulf of Mexico were mainly plains that were submerged during a seismic catastrophe twelve thousand years ago. Only the highest peaks remained above water and these became the Caribbean islands of today.

Some reports state that up until ten thousand BC there was a large icefree tundra plain between Eastern Siberia and Alaska where the Bering Strait is now. This was possibly because so much water had been locked up in the vast Polar Ice Caps. Other sources state that Beringia disappeared under water five hundred years before but our dates are being found to be variable. I would allow for five hundred to one thousand years when dates are this close to flatten out the geological dating variables. Do you see how we need to combine the dating of the reports?

Some scientists theorize that the islands of Bermuda were the peaks of immense underwater mountains that were above sealevel during the last Ice Age. These islands were more than likely the central uplifts of the impact crater. What is a central uplift? When a meteorite strikes the earth the surface of the earth is melted and a splash occurs in the middle of the crater. Depending on the size of the impact is the size of the central impact. The central uplift is molten material that solidified as it headed back to earth whilst cooling. You would have seen similar things in videos of milk producing the same effect when something is splashed into it.

There are oriented shallow depressions or lakes near Beni in north-eastern Bolivia that are aligned northwest to southwest. These are the same age and the same direction and shape as the Carolina Bays and are from the same period. These lakes cover 45,000 square kilometers. In all of these lakes one rim is more developed than the others. This is the same as around Point Barrow in Alaska and the contemporary Carolina Bays. One end of these oval lakes has barely any edge deposits yet the opposite edge has a definite high rim as if something skidded at a low angle into the ground and created a massive divot. Falling comets or ice masses coming in at low angles could do this and they have been theorized as occurring en masse around this time. Comets or falling ice masses would have left very little traces.

The Araona meteor crater was discovered in the Amazon jungles of Bolivia in 2002 by the NASA space shuttle. The crater is five miles wide and only sixty-six feet deep and very similar to the Carolina Bays of the United States which are also wide but shallow. It appears to be a cometary impact, possibly an ice ball. When an iceball or cometary fragment hits the Earth there would be an impact crater but no meteor fragments as the mass is primarily mud and ice which would shatter on impact after creating the impact crater. Is the Iturralde Crater from the same shower as the meteorites that created the shallow lakes nearby in Beni? The Araona Crater is in Iturralde in La Paz in Bolivia.

Now we come to Tiahuanaco or Tiwanaku on the shores of Lake Titicaca in La Paz province in Bolivia. There are no elephants in the Americas though there had been in prehistoric times. They were particularly numerous in the southern Andes until their sudden extinction around 10,000 BC. They had been members of a genus of elephant called *cuvieronius* uncannily similar to the elephants depicted on the Gate of the Sun at Tiahuanaco. There were other extinct species as well carved on the gate such as the *toxodon*, a three toed amphibious mammal about nine feet long and five feet high at the shoulder resembling a short stubby cross between a rhinoceros and a hippo that had died out 12,000 years ago at the end of the Pleistocene Era. There were no fewer than 46 *toxodon* heads carved into the frieze of the Gateway of the Sun. Were these mistaken for elephant heads by some researchers? There were also numerous fragments of pottery in the area with *toxodons* painted on them as well as in several three-dimensional pieces of sculpture. Other species found were the *shelidoterium*, a diurnal quadruped, and *macrauchenia*, an animal larger than the modern horse with distinctive three toed feet.

    The Altiplano of Bolivia suffered great floods in the Andes Mountains during the eleventh millennium BC. How do you flood the tops of mountains? How do you get great floods on the tops of very high mountains unless the mountains were much lower and suddenly raised themselves to their present day heights? In the ranges of the Andes at an altitude of 13,000 feet geologists have found stretches of marine sediment reaching 640 kilometres all the way from Peru to Bolivia, clear evidence that the level of the ocean only some tens of thousands of years ago was 13,000 feet higher than today. Similar sediments have been found in the Himalayas. Were these landmasses in fact much lower? Was the rising much faster and more recent? At a height of 11,500 feet there is a curious white streak running along the side of the Andes for over 300 miles. It consists of the calcified remains of marine plants indicating that the slopes in question were once part of the seashore. These are thousands of years old. Some researchers suggest that Tiahuanaco was originally built at sea level and that it was thrust up over two miles in altitude during a convulsion of the Earth. This theory is based on the discovery of the watermark line on the surrounding mountains that stretches for over 300 miles and consists of calcified remains of marine plants. Lake Titicaca has a very high saline level and oceanic fauna, as do other lakes in the area. There are also the ruins of what appear to be a ruined seaport close to the city. Some German and local archaeologists believe that the city was abandoned around 10,000 BC to 9,000 BC. Others say that Tiahuanaco is less than two thousand years old. Irrespective of this the fact that Lake Titicaca shows that it was raised and then tilted. When and how?

    There is a freshwater seahorse living in Lake Titicaca. It is the only known freshwater seahorse and certainly the only seahorse living at an altitude of 13,000 feet. Lake Titicaca is a saltwater lake though so the seahorses may not technically be freshwater ones but saltwater ones. How did the seahorses get up to the altitude that they are now? Lake Titicaca must have been connected to the sea once for the seahorses to actually enter it. Was this before it was thrust up into the sky? Oceanic mollusk shells have also been found here.Speaking of oceanic fauna in odd places? Corals attached to rocks underlying the nitrate deposits along Lake Titicaca's waterless western coast are the same as existing species of corals on the Peruvian coast. These corals of Lake Titicaca are now at very high altitudes above the present sea level.

    Lake Titicaca is one hundred and ten miles long and thirty five miles deep with a maximum depth of 890 feet. It is a salt water lake now at an altitude of 12,300 feet.

The area around Lake Titicaca was well wooded until comparatively recently and had a rich and varied local fauna. Trees cannot grow at the present altitude. Incidentally ancient legends state that the gods who brought agriculture to Tiahuanaco were said to have come out of the regions of the south immediately after the deluge. According to the same legends the area then was covered with forests and thickets. This indicates that it was at a lower altitude. Changes in the Andean Ridge as interpreted by deposits of calcareous lime or watermark lines on cliffs or mountains indicate that the area was thrust upwards. This line is quite visible and noticeably slanting as if the shoreline tilted. Remains of salt lines in the mountains may indicate that the city was not a mountain fastness but possibly an ocean port. The Russian scientist Alexandre Kazantzev wrote that Lake Titicaca was lifted to its present altitude by a natural cataclysm and was formerly an inlet of the sea. The old shoreline is still visible as well as shells and other marine plants that confirm that this was originally sea. There are also the remains of an apparent seaport here constructed of cyclopean blocks of stone. This could be quite recent in fact and has no bearing on the age of the lake's uplift.

Remains of mastodons, toxodon and giant sloths, which could not live at the present altitude, have also been found here. Pictures of the same animals have been found on ceramics found here as well.

The Philippines, Sumatra, Borneo and Java were together connected to the landmass of continental Asia up until the Cataclysm of 10,000 BC. This geographic area was called Sundaland.

Professor Gregory reasoned that the old highlands of Brazil must have extended eastward into the Atlantic and were built largely of materials derived from the destruction of an old Atlantic land.

There is postulation that immense earthquakes around 10,000 BC drained the inland Amazon Sea in Brazil into swamps that later became jungles.

Along the Arctic coast of Canada there is a marine deposit containing walrus, seals and at least five genera of whales overlying the seaboard. The same remains are along the northeastern states of Canada as well.

There is a 150 mile or 240 kilometre wide submerged meteor or cometary crater in the Amundsen Gulf in Canada that is from the period of 10,000 BC.

There is also a 75 mile or 120 kilometres wide submerged meteor or cometary crater off Baffin Bay in Canada. Are these more meteor or cometary craters from this period? How large was the rain of stones?

The Bloody Creek Crater in Newfoundland in Canada is a small crater in granite that was formed by an impact twelve thousand years ago. There are several other smaller craters around it from the same period. Bloody Creek Crater is elliptical in shape and is four hundred and twenty metres on its long side. The site is now submerged under a hydro-electric power project. Bloody Creek Crater is near Scrag Lake in Nova Scotia in Canada.

North of Fairbanks, Alaska and in the Yukon Valley as well as around North Star in Alberta, Canada, deep frozen woolly nammoth remains have been taken from deep in the ground during the extraction of gold with high-pressure pumps and excavators. The deep frozen stomachs contain leaves and grass, which the animals had eaten. The young lay next to the old, the babies beside their mothers. Such quantities of animals cannot have died all at once in a natural way. The animals had died almost instantaneously and were deep frozen on the spot otherwise they would have shown minimal signs of decomposition. In addition 1,766 jawbones and 4,838 metatarsal bones belonging to a single species of bison were found near Fairbanks. The mammoths were corpses before they were frozen. It is only an assumption that they were hairy so as to survive in frozen climates. Cats have fur and survive in the tropics today.

The mean temperature of Baffin Island in this period was warmer than it is today when the North Pole was then at Hudson Bay. This was realized by examination of alder and birch tree remains in peat bogs which are composed of decayed vegetable matter. This indicates that Baffin Island was ice free in the last Ice Age. This was while the Wisconsin Glaciation in North America was expanding. The warmer Arctic Sea would have heated up Baffin Island though it was closer to the then North Pole than it is now.

The bones of a whale have been found north of Lake Ontario about 440 feet above sea level as well as in the Montreal-Quebec area about six hundred feet above sea level.

Stratigraphic core samples from Newfoundland contained layers of volcanic ash dating from the transition period from the Quaternary to the Quinternary age, the period of the last glaciation, 10,000 BC. What sort of Hell was this with meteorites and ice masses crashing down from the skies as well as volcanic eruptions, earthquakes and tidal waves?

Here you can see the enormous Bermuda Crater and northeast of it the Hudson Bay Crater and then Amundsen Gulf and then Point Barrow. Smaller craters are also indicated. The nearest selection to the Bermuda crater are probably fragments falling as the Earth turns to the east. Large fragments crash in a row to the northwest. This formation of craters is called a tear drop.

While Glaciers were grinding deep into North America during the last Ice Age, the Yukon and some Arctic islands that are now covered in ice were ice-free.

The sea receded away from China to a huge extent and at the same time massive tidal waves hit Ancon in Peru. This was the tsunami of the millennia during the Cataclysm of 10,000 BC.

Now back to normal again. If you can call it that? After having occupied the English caves for untold ages Paleolithic man disappeared forever and along with him many animals that are now either locally or wholly extinct. Above the remains of man in these caves comes a deposit of stalagmites, twelve feet thick, indicating a vast period of time during which it was being formed and during which man was absent. Above this stalagmite comes another deposit of cave-earth. The deposits immediately overlaying the stalagmite and cave earth contain an almost totally different assemblage of animal remains along with relics of the Neolithic, Bronze, Iron and historic periods. There is no passage but a sharp and abrupt break between the later deposits and the underlying Paleolithic remains.

Around 10,000 BC the giant rhinoceros of Europe became extinct.

There are distinct curving lines visible from the air at Lake Saimaa in southern Finland that show a circle with a 150 mile long arc enclosing a host of shallow lakes that was formed at the end of the last Ice Age. The edges are formed by glacial till composed of sand and gravel pushed up by glaciers like bulldozers. This is believed by Richard Firestone, Allen West and Simon Warwick-Smith to be a shallow cometary or ice crater similar to the Carolina Bays in the United States formed in this same period.

In the Cave of Le Mas d'Azil in Ariege in the Midi-Pyrenees of France sixteen miles northwest of Foix there are painted signs on pebbles found here in the cave that are possibly writing and are 12,000 years old, 10,000 BC, the period of transition between the old Stone Age and the New Stone Age. The stones are almost all oval shaped, ringed by a line of paint and decorated on one side with an abstract sign such as a cross, a circle or a line and so on. Many resemble ancient types of written symbols. The French prehistorian Edouard Piette discovered the stones at the end of the Nineteenth Century. One of the most interesting features of Azilian art was a collection of pebbles found at

Mas-d'Azil in the Ariege that were painted red with a mixture of peroxide of iron and some resinous substance and displaying zigzags, crosses, circles, vertical strokes, and ladder like patterns resembling the letter E. This site is where the remains of the Azilians were found. The Azilians were apparently vegetarian or fruitarian and the presence of barley seeds indicated that they also cultivated cereals. Harpoons were also found indicating a maritime origin. They might not have been as vegan as some would want us to believe. Fish aren't vegetables or fruits. The Azilians are named after the cave of Mas d'Azil where their remains were first found.

    The Azilians arrived in Europe around 10,000 BC. Their handling of flints and bones was of the highest delicacy and they had high geometric art which was very symbolistic and could be a developmental stage to writing. The men wore feather headdresses and short trousers and the women wore short skirts and caps covered with ornaments. Azilian culture is found in North Africa and southwest Europe. Feathered headdresses were also popular in the Americas. The Azilians buried their dead with their faces facing west. The Azilians incidentally were a sea-faring race but where did they learn the art of navigation? The Azilian flint fish-hooks are too big except for deep sea fishing and they were regarded as a population of fishermen. Incidentally there were many Azilian sites in the Bay of Biscay. The same area used as the arrival point for the *Cro-Magnon* people incidentally. Numerous extremely realistic carvings have been found here as well. Stylizations on the head of a horse indicate that halters were in use here thousands of years before their invention. Except that we have already had sculptures of horses wearing halters dating from before this period.

    Found near Mazerolles in Rochebertier in the Charente of France is a twelve thousand year old reindeer bone that has markings on it that appear to be more than just decoration. They appear to be letters or some form of writing. The writing is similar to the Tartessian script of Iberian Spain that is supposed to be six thousand years later.

    Another report stated that there are repetitive marks in a cave at Rochebertier in France that may be picture writing, a tally or even an alphabet. The marks though are 8,000 to 10,000 years old. Isn't this the same period as the marks found at Mas d'Azil, which is near Foiz? Once again a disparity in dates and isn't this always the case? The oldest accepted writing is from Sumeria in the Middle East from around 2,700 BC yet we are seeing what are possibly older writing styles.

    At the close of the Tertiary Period there were forests of beeches, maples, walnuts, poplars, pine trees, oaks, sequoias and magnolias in Greenland and Spitzbergen. The remains of these trees are found jumbled together as if something pushed them over and then bulldozed them but did not crush them. Glaciers would have crushed them. These trees were pushed over and not crushed. Tsunamis can do this. Drift has been found here that is thirty feet above sea level that is woven with whale bones. What pushed the whales up onto the land? More tidal waves? Now you see why we might have to combine a lot of the dates for the data as it does get confusing with the different dating systems. All the phenomena are the same or similar but the dates are all over the place over the three thousand year period.

At Atane-Kerdluk in northern Greenland at latitude 70 degrees north at an elevation of more than one thousand feet above the sea were found the remains of beeches, oaks, pines, poplars, maples, walnuts, magnolias, limes and vines. There are no trees here now nor have there been for a long time.

Further developments in research in the Gulf of Mexico show from the study of fossil shells of *Foraminifera*, a marine life form, that between eleven to twelve thousand years ago, 10,000 to 9,000 BC, there was a substantial increase of twenty per cent in the warmth and a twenty per cent decrease in the salinity of the water. This indicates that there was a profound change in the level and the temperatures of the oceans as glaciers started to melt at this time. This also allowed for more freshwater melt pouring into the North Atlantic and reducing the salinity levels. There are two types of Foraminifera. The two principal species are *globorotalia menardii* and *globorotalia truncatulinoides*. The first is distinguished by a shell spiraling left wise and lives in warm water. The other spirals right wise and lives in cold water. The warm type does not appear anywhere above the line stretching from the Azores to the Canaries. The coldwater *Foraminifera* are present in the northeastern quadrant of the Atlantic. The warm type inhabits the middle Atlantic zone from West Africa to Central America. Yet in the equatorial Atlantic the cold type show up again. It looks as if the warm species of *Foraminifera* tore through some barrier and headed in an easterly direction. Based on *Foraminifera* distribution the Lamont Geological Observatory discovered that a sudden warming of surface ocean waters occurred in the Atlantic about 10,000 years ago. The transformation of the cold type foraminifera to the warm type did not last more than one hundred years.

The inhabitants of Zawi Chemi used grinding stones to produce flour from wild cereals as early as the late tenth Millennium BC around 10,000 BC. This site is near the Shanidar Cave Site in Kurdistan in Iraq which we have met before..

In Ireland the remains of trees have been found in sand beds under the glacial till of the last Ice Age.

In 10,000 BC Ireland was Arctic and covered in ice. It was a scene of Arctic desolation. Today the Gulf Stream and the Irish Sea provides subtropical vegetation and palm trees on places. There was no Gulf Stream in this period. The Gulf Stream seems to come and go.

Pottery was not supposed to exist twelve thousand years ago. Who then made pottery near Nagano in Honshu in Japan? Pottery jars were in use here around ten thousand BC. Other pots found on Honshu were one thousand years older from 11,000 BC. Pottery was not supposed to have been invented yet. It seems that the Japanese did not know this. Other sources state that this was around 16,000 BC. The pots were found in Ishigoya Cave.

Around twelve thousand years ago, 10,000 BC, the Mediterranean Valley was flooded by the Atlantic Ocean, in a period when world sea levels had risen by two hundred metres or six hundred feet. Some cosmic event had caused the northern Polar Ice Cap which covered much of North America to suddenly melt and raised the levels of all the oceans and caused the Atlantic Ocean to rush through into the Mediterranean Sea at the Strait of Gibraltar..

Yes, I know that there have been several dates for this event, so this is another one. What is not in dispute is that it actually happened.

Tools and weapons have been found in Morocco that weighed ten to twenty-two pounds and could have only been used by giants. The Berbers of Morocco are a light-skinned people who are believed to have arrived in North Africa from the Middle East around 10,000 BC. They are called *Capsian Man* after discoveries at ancient Capsa, modern Gafsa, in Tunisia. The artifacts are of well-made and unusually large stone axes. Did they arrive from the Middle East though? Their earliest legends state that they came from a lost island in the Atlantic. Who do you believe?

The continental ice sheets started to expand again in Norway sending long tongues of glacial ice down mountain valleys and out into the fjords.

In the Pacific Ocean there are gujots which are flat topped seamounts that were above sea level prior to 10,000 BC. Now their flat tops are all at the same level under the sea. They were once whole archipelagoes with the gujots being the tops of hills and mountains. Their tops are all at the same depth of three hundred to five hundred metres below the surface yet they were above sea level at the end of the last Ice Age. Coral reefs could not grow fast enough to keep up with the sudden submergences.

The Nazca Ridge in the Pacific Ocean of the west coast of South America is crowned with submerged gujots. The Nazca Ridge runs from Easter Island to the South American Coast. The ridge itself is now at a depth of one to two kilometres whilst the tops of the gujots are between two hundred and five hundred metres down.

Now for another archaeological diversion. In 1927 AD there was a discovery by the American Indian Foundation Fund at Cocle under many feet of volcanic ash. This was of a great temple, coloured pottery, stone monuments and idols or statues of men. They are believed to be 12,000 years old. This was at Cocle in Panama. The American Archaeologist A. Hyatt Verrill believes that some of the Cocle ceramics depict a flying lizard like a *pterodactyl*. There were many of these ceramics that depict in perfect detail the flying *saurians* such as the *pteranodon* and the *pteradactyl*. We shall meet more of these flying reptiles as we go on. Also in the Cocle pottery was a picture of an elephant with a long trunk, ears like great-fringed leaves and a saddle on its back. The American elephant is often thought to have become extinct in this period though other sources state that the elephant survived until 8,000 BC. Other sources state even later and a few maverick scientists say that they are still around in the more remote areas of North America.

The Persian Gulf was completely dry until 12,000 years ago and was a refuge from the Ice Age World. In the late Ice Age the Persian Gulf was still dry land and global sealevels were ninety metres lower than the early twenty-first century AD. The Tigris and Euphrates Rivers flowed through deep valleys into the Gulf of Oman eight hundred kilometers south of their present estuaries. As sealevels rose during the Great Warming the newly formed Persian Gulf caused massive alluvial buildup in the Mesopotamian Plain where the gradient was extremely low with a mere thirty metre drop over seven hundred kilometers. This resulted in sluggish rivers that flowed slowly and marshes and swamps were everywhere. The courses of the two rivers changed frequently. Was this now submerged land area the mythical Eden? After all the Tigris and Euphrates Rivers are mentioned as flowing into Eden? Were the two other rivers that flowed into Eden, the Pishon and Gihon, rivers that now flowed into this now submerged once lush area? The Persian Gulf, also known as the Arabian Gulf, was a large plain area that rapidly became submerged as world sea levels suddenly rose. Was this where Eden was? We have a lot of candidates for Eden. The Persian Gulf and also the Euxene Lake are both candidates.

Lake Racze on Wolin Island in Poland is an impact crater that is three hundred and twenty metres in diameter and twelve thousand years old. There are impact furrows on the eastern side of the lake which is one of a dozen small lakes forming a chain in the centre of the island. Possibly the other lakes are impact craters as well. Racze Lake is near Kolczewo.

So much was happening around 10,000 BC it was unbelievable.

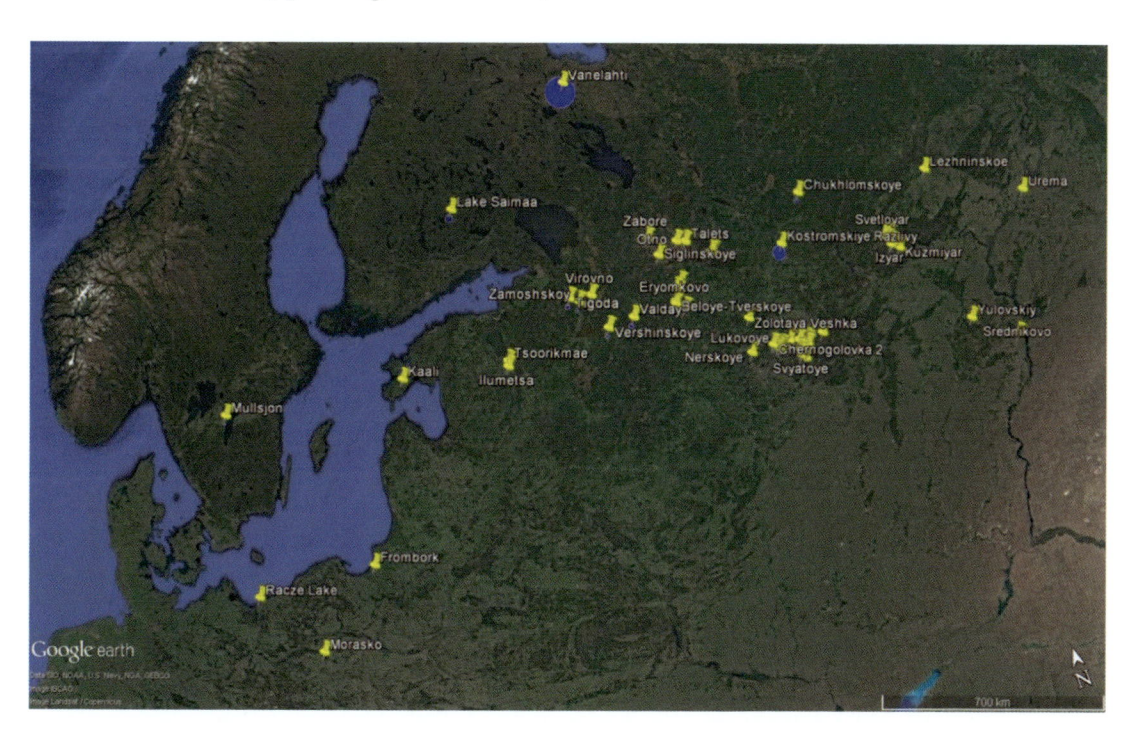

These are the impacts east of the Russian impacts of ten thousand years ago. They are all from the period of ten thousand BC to 8,000 BC. Are they all separate impacts at different times or part of one meteoric Megastream?

The continental ice sheets started to expand again in Scotland sending long tongues of glacial ice down mountain valleys and out into the firths.

The bison of Siberia suddenly became extinct along with almost everything else including mammoths, mastodons and woolly rhinoceroses.

At a site on the Liahknov islands in the far northern Pacific region known as the Laptev Sea a 12,000-year-old deposit was discovered in 1905-1906. This deposit was composed of millions of broken trees two hundred feet deep. One violent blow of nature had wrenched them all out, roots and all. Encased in the mud beneath the splintered wood were the remains of the broken bodies of millions of animals such as the mammoths and bison. Ivan Lyakhov, the discoverer of the islands stated that from viewing the enormous quantities of mammoth remains there that it seemed that the island was composed almost completely of bones and tusks. For a time the area supplied half of the world's supply of ivory for piano keys and billiard balls. Undigested foods from plants that currently do not grow in the area for hundreds of miles around were discovered in the animals' stomachs. Fractured bones and other evidence indicate a sudden, violent, shock. Cuvier theorized that a huge wave had swept these creatures up from some distant spot and dragged them through what was then a forest tearing them apart as well as burying them. All over the island the bones and tusks were found of elephants, as well as the bones of rhinoceroses, buffaloes and horses in such vast numbers that it was thought that these animals must have once lived here in enormous herds. Whole plum trees have been found which nowadays cannot grow within two thousand miles from the North Pole. They could not have been buried in frozen muck which is rock hard and would not have retained their fruit and leaves if they were washed up from where they had been growing. This indicates that the Earth has tilted and in the period that the trees were last growing there was long enough sunlight as well as heat for them to fruit and survive. Liakoff Island is in the Liahknov Islands in the New Siberian Islands.

The broken remains of a sunken city were found on the seabed near Cadiz, formerly known as the Phoenician port of Gades, in Andalucia in Spain. The ruins of a port complex covering an extensive area have been found in the sea here and the last time that the area was above sea level was around 9,000-10,000 BC. The ruins include roads and large columns, some with concentric spiral motifs the same as in Malta and Ireland. The legend of Cadiz is that a long time ago the Pillars of Hercules had split open and the Mediterranean had begun to flood. At the ancient temples sacrificial offerings were made to the Gods but it was to no avail. After the Cataclysm Cadiz was founded and new temples to Baal, Astarte-Venus and Melkart were built. Cadiz was an early Phoenician Atlantic port with an excellent harbour. In Phoenician times the city was divided into different sections. In the west part, Punta del Nao, was the temple of Astarte-Venus, Venus, and to the east on the San Sebastian causeway was the temple of Baal while the Phoenician city lay inland. To the far eastern end of the Cadiz peninsula was the temple of Melkart, Melgart, Hercules. Today the temple of Baal and the temple of Astarte-Venus are submerged megalithic ruins. The temple of Astarte-Venus is submerged in the bay off the Santa Catalina area. The massive ruins of the temple of Baal can be viewed in the water near the lighthouse on the San Sebastian causeway. The giant blocks of stone weighing 40 to 80 tons lie in about eight feet of water. The square stones have slots and grooves in them to be fitted with other stones.

Another mysterious ruin has been found at Huelva in Spain. If we really want to find the remains or traces of cities from before the period of 9-10,000 BC we must look at where our greatest cities generally are now for we shall find that the ancients had the same reasons for siting their cities as we have for ours. This was for the best trade or defensive position or preferably both for the future growth of the city and its security. The majority of our cities are on the coast or have access to major waterways for then as now the largest goods shipments are by water. Therefore we must look not at where the shoreline and the best bays and anchorages are now but at the precataclysic shoreline and where the best bays and anchorages were. There are enormous ruins in the sea here similar to nearby Nielva that appear to have been parts of ports, docks and seawalls. There are also stone staircases on the seashelf which was above sealevel prior to 9-10,000 BC.

The Philippines, Sumatra, Borneo and Java were together connected to the landmass of continental Asia up until the Cataclysm of 10,000 BC. This geographic area was called Sundaland.

Around 12,000 years ago as the sea levels started rapidly rising a continental expanse began to sink beneath the waves in Southeast Asia. This was the Sunda Shelf or Sundaland and it lay where the southern reach of the South China Sea, the Gulf of Thailand and the Java Sea are all now. A land area equivalent to the Indian subcontinent vanished below the sea leaving only the Malay Archipelago, Indonesia, Indochina and Borneo protruding from it. A large section of land along the Pacific coast of East and Southeast Asia was submerged as well. At present these seas are only several hundred feet deep. Originally this area was probably inhabited but nothing is known of who or where except the dimmest of legends.

The Swiss scholar Francois Forel calculated that the quantity of detritus and its velocity carried by the Rhone River to Lake Geneva in Vaud in Switzerland from the Alps showed that the last Glacial Period finished only twelve thousand years ago.

According to some scientists the Swiss Alps were only three thousand feet tall in this period before they were suddenly raised to their present heights. The Matterhorn and Mont Blanc are now over 14,000 feet high. Immense walls of stone thousands of feet high and miles wide were suddenly thrust up.

An unusual item was found at Mezhirich mear Cerkaska in the Ukraine. A map was found here in 1966 that was engraved on a mammoth tusk. The map shows a local river flanked by a row of houses and was dated to ten thousand BC Mezhirich is famed as a mammoth village. There were the remains of four huts made from 149 mammoth bones dating back 15,000 years, 13,000 BC. A farmer, digging his cellar, almost two meters below ground level, struck the massive lower jaw of a mammoth with his spade. The jawbone was upside down, and had been inserted into the bottom of another jaw like a child's building brick. In fact, as subsequent excavation showed, a complete ring of these inverted interlocking jaws formed the solid base of a roughly circular hut about four or five meters across. About three dozen huge, curving mammoth tusks had been used as arching supports for the roof and for the porch, some of them still left in their sockets in the skulls. Separate lengths of tusks were even linked in laces by a hollow sleeve of ivory that fitted over the join. It has been estimated that the total of bones incorporated in the structure must have belonged to a minimum of ninety-five mammoths. A mammoth skull was also found here that has red painting on the front of the skull of unknown use. The mammoth skull was found at the front of one of the huts.

Lake Aksehir on the Konya Plateau in the west of Anatolia in Turkey was five times its present size as recently as thirteen thousand years ago. During the Younger Dryas it had shrunk down to a pond surrounded by cobbles and gravel that were high and dry.

The turmoil that caused the extinction of so many species of animals in both the New and Old Worlds according to Charles Darwin must have shaken the whole framework of the Earth. In the New World, more then seventy genera of large mammals suddenly became extinct between 15,000 BC and 7,000 BC. This included all North American members of seven families and one complete order, *proboscidea* or elephants. These staggering losses involving the violent obliteration of more than forty million animals were not spread out evenly over the entire period but were in a small period of two thousand years between 11,000 BC and 9,000 BC. In perspective in the previous three hundred thousand years, only about twenty genera had disappeared. This pattern was repeated across Europe and Asia. Even far off Australia was not exempt losing around nineteen genera of large vertebrates, not all mammals, in this relatively short period.

There were elephants roaming the plains of North America and South America prior to twelve thousand years ago, 10,000 BC, when they suddenly became extinct for no apparent reason. Plato said that elephants also roamed the plains of Atlantis. Elephants had also roamed the plains of Europe as well.

Paleontologist George G. Simpson considers the extinction of the Pleistocene horse in North America to be one of the most mysterious episodes in zoological history. The horse was not the only extinction in this period. Giant tortoises living in the Caribbean, the giant sloth, the sabre-toothed tiger, the *glyptodont* and *toxodon*, all tropical animals, all became extinct. Woolly mammoths, woolly rhinoceroses, giant armadillos, giant beavers, giant jaguars, giant ground sloths and scores of other complete species were wiped out at the end of the Pleistocene. This was the end of the age of the giant mammals.

The bulk of the animal extinctions took place between eleven thousand BC and eight thousand BC when there were violent and unexplained fluctuations of climate. Geologist John Imbrie stated that there was a climatic revolution around eleven thousand years ago, 9,000 BC. There were also increased rates of sedimentation and an abrupt temperature increase of six to ten degrees centigrade in the surface waters of the Atlantic Ocean. This increase in water temperatures alone would have brought about the end of the Ice Ages.

All over North and South America Ice Age fossils have been unearthed in which incongruous animal types, carnivores and herbivores, are mixed promiscuously with human bones. There are also widespread areas of fossilized land and sea creatures mingled in no order and yet entombed in the same geological strata. It was as if massive tidal waves had overtaken them! Again!

In the United States Ice Age marine features are present along the Gulf Coast east of the Mississippi River, in some places at altitudes of two hundred feet.

Innumerable Clovis era flint spear points found across North America are peppered with micrometeorites that resemble tiny black magnetic grains, even leaving microcraters and impact trails in the hard flint indicating tremendous velocity and impact. Numerous megafuana, mammoth and mastodon, bones are also pitted with these micrometeorites, always on one side only the same as the Clovis Points that have been discovered from this period. The angle of impact of these micrometeorites levels out the further away that you go from the Great Lakes indicating that they were falling straight down over the Great Lakes and at increasing angles away from the Great Lakes as the number of micrometeorites and particles decreases. The micrometeorites were tiny iron spherules and cannot be created on Earth. Volcanic spherules are glass and not iron.

Another anomaly is the "black mat" which is always found above Clovis era sites and the occurrence of the iron micrometeorites. The "black mat" was probably composed of algae that experienced an explosion of growth in North America at the end of the Clovis Era. In many cases the "black mat" covers the remains of extinct animals from this same period suggesting that the "black mat" occurred after the extinctions.

At the end of the Bolling Phase and the return of the extreme cold the climate became considerably wetter with long and frequent soaking rain that covered everything and these lakes, marshes and ponds began to be quickly covered with thick choking mats of blue-green algae. As the algae died they sank to the bottom and covered the remains of the megafauna and the Clovis era spear points. The blue algae could have also killed numerous animals as it is quite toxic. Blue algae or blue-green algae is called a cyanobacteria that has the ability to produce toxins. There are still arguments as to whether it is an algae or a bacteria but irrespective of what it is it can kill animals and humans. One theory is that the "black mat" grew so rapidly due to the presence of the iron micrometeorites that had just fallen. The conditions would have been ideal and the iron meteorites would have caused an algal bloom which was quite poisonous as it still is today. Had the toxic "black mat" killed off Clovis Man and the megafauna? The "black mat" is found in the United States, Canada and Mexico and often drapes over the skeletons of the dead megafauna. Quite often above the "black mat" there are numerous occurrences of charcoal from fierce fires.

In North America the Clovis Era cultures all disappear at the same time. Incidentally the only other culture to practice the same flaking techniques as the Clovis site spear point makers were the Solutreans from Northern France and Spain. No one in Siberia created Clovis Points yet the Clovis Point culture by accepted reasoning is supposed to have originated in Siberia and travelled across Beringia, now the Bering Strait, into North America. Specialists cannot usually tell the difference between Clovis and Solutrean spear points though the spear points are easily distinguished from any other type of spear point. The Solutrean spear point culture ended around 14,000 BC before Clovis appeared in North America around 11,500 years ago. Allowing for the fact that water travel was the accepted means in early history were there colonists from Europe residing in North America in this period?

There are great meteorite craters found in the Carolinas, Georgia, Florida and Virginia dating from ten to twelve thousand years ago, 10,000 to 8,000 BC. It is always hard when you have such large variance in dates. Also remember that geological dating variance is quite common.

The Carolina Bays for example are elliptical and average half a mile in diameter with elevated rims and depressions 25 to 50 feet deep. The northwestern ends have low rims and the southeastern rims are higher as if something skidded at an angle in that direction. The Carolina Bays have also been called the Carolina Craterfield. In 1931 the authorities in the two adjacent states of North Carolina and South Carolina decide to try out a new photogrammetric survey and commissioned a company specializing in aerial photography to undertake the task. The stretch between Florida and Cape Hatteras and its hinterland were exposed to the aerial camera. When the films were enlarged and examined in the stereo-comparator they created a sensation.

The pilots who saw the pictures were startlingly reminded of the battlefields of the 1914-1918 War. Huge circular or oval shapes sometimes overlapping that could be clearly identified as enlarged mud filled craters caused by the impact of gigantic boulders characterized all the pictures. The existence of the Bays had been completely unknown in the flat marshy land that was traversed by troughs or bays. The origin of the bays was now suspected. There were three thousand such troughs. Half of them are longer than 1,300 feet and more than one hundred are longer than 5,250 feet. They cover an elongated ellipse of 63,500 miles. Only a small portion of this is on land. The rest is on the fragmented coastal strip. All the longitudinal axes are parallel indicating that the fragments must have consisted of boulders traveling parallel to each other. Even though the craters are 11,000 years old most of the troughs still display remains of the thrust wall in the Southeast. The boulders must have come from the Northwest. The cause of the craters or bays was attributed to the Carolina Meteorites. Because the bays are elliptical as well as circular indicates that the descents and angles of impact must have differed considerably. Only a very large and disintegrating celestial body could have done this. A meteorite shower, if one were large enough would have produced identical holes, either circular or oval, not mixed. A descending and exploding large cosmic body could only produce the Carolina Bays. When the nucleus is traveling ahead of the tail of fragments the craters overlap. Another theory is that the bays were formed by masses of ice ejected from the area of the Great Lakes and that is why no meteoric masses have found in them. In some of the bays on the low rims for the bays are shallow and elliptical generally with their end points parallel to each other, meteoric black glass fragments have been found that are the same as the meteoric glass fragments found in Clovis sites before their disappearence. The sand around the bay rims is white as well and this indicates massive exposure to heat as the sand was blasted away from the impact zone. This rules out ice masses from the Great Lakes and sounds more like cometary fragments. There are over one hundred thousand meteor craters concentrated along the Georgia, Florida, Virginia and North and South Carolina coasts but also extending North and South indicating a great meteor swarm once struck the area.

At the end of the Ice Age the massive volumes of waters suddenly draining away from the collapsing and thawing ice masses would have swept down river valleys across North America at the same height as todays skyscrapers. The flow rates were over one hundred times the flow rates of rivers today and many lakes were formed that were larger than many modern American states. As icedams collapsed massive surges of water would have acted as giant bulldozers across the land sweeping all before them in their path.

The Mississippi River had suddenly backed up and instead of dumping silt and clay into the Gulf of Mexico sent it backwards where it was retained in the alluvial valley and delta.

The mastodons of Alaska suddenly became extinct. Along with a lot of other gigantic and smaller animals that appeared to have been mangled and dislocated by enormous tidal waves. Tidal waves coming from inland! From rapidly melting glaciers and collapsing ice dams!

Whilst miners were mining for gold in the Goldstream Valley of Alaska they uncovered thousands of trees still standing in their original positions but with the rest of the trees sheared off six feet above their bases. Then it was all buried. It was as if a giant axe had lopped the trees. They did not appear to have been crushed by glaciers but snapped by a sudden onrush of matter and water.

There are oriented shallow depressions or lakes in the permafrost near Point Barrow in Alaska that are aligned northwest to southeast. These are the same age and the same direction and shape as the Carolina Bays and are from the same period. Was it possible that during the meteorite bombardment there was a separate bombardment in North America that smashed the glacial field which exploded showering ice in all directions? Or were these cometary fragments?

The Sithylemenkat Crater near Bettles in Alaska is twelve and a quarter kilometres in diameter and is twelve thousand years old. It is just south of Bettles and is occupied by Sithylemenkat Lake. The crater is five hundred feet deep. There is an abnormal proportion of nickel inside the crater as well as around its edges in the peripheral impact zone.

Archeological excavations at Windmill Lake in Alaska show that the lake bottom was catching beetles common to much warmer climates with summer temperatures similar to the present day. Only one thousand years before the beetles were those found in the Arctic tundra. This indicates a massive temperature drop.

To the southeast of Painted Desert in Arizona is the Great Meteor Crater or Winslow Meteor Crater, also called the Barringer Meteor Crater. Next to it is the amazing Petrified Forest. Approximately 20,000 to 12,000 years ago, 18,000 BC to 10,000 BC, a meteor crashed here creating a huge crater. This was at the same time as the Odessa group of meteorite craters in Texas. The rain of meteors in Arizona covered 550 square miles. The meteors struck out of the northern sky, weighing one million tons and traveling at several miles per second. The Barringer Meteor struck at an angle driving through solid rock while it was being decelerated to zero velocity and its fragments finally came to rest at the base of a cliff. Half a billion tons of rock were displaced. The crystal structure of the meteorite fragments indicates that it might be an Asteroid, the remains of an exploded Planet orbiting between Jupiter and Mars. From the ring to the pit bottom the depth is 570 feet. The diameter of the top of the rim is 4,000 feet. The circumference is three miles. The meteorite might have been an iron sphere 60 feet in diameter. The meteorite made a hole 4,500 feet across and 600 feet deep. It flung out masses of rock weighing up to 7,000 tons and altogether hurled out 400 million tons of rock. The pressure of the impact exceeded one million pounds per square inch turning silica into new forms known as coesite and stishovite which are both indicators of impact events. Another date for the creation of the impact crater is fifty thousand years. Take your pick.

The Navajo have a legend that one of their sky Gods streaked down from Heaven as a flaming serpent one day long ago and blasted out the bowl of Flagstaff Meteor Crater as a token of his passing. The Navajo were not around fifty thousand years ago so the odds are that they witnessed the cosmic impact twelve thousand years ago. What then of the Barringer meteor shower in Arizona around this same time as well as the Odessa Meteor shower in Texas which was contemporary with it?

Several authorities state that we had the Barringer meteor crash in Arizona around this same time as well as the Odessa Meteor shower in Texas which was contemporary with it? The Odessa Meteor Shower struck an area of Ector County in Texas. The Odessa Crater Field has at least four minor craters as well as a main crater. Over one thousand five hundred meteorite fragments have been found here weighing up to three hundred pounds. The largest crater is one hundred and sixty-eight metres in diameter and exposed to the surface. It was originally one hundred metres deep. Around Odessa and Midland in Texas there are hundreds of salinas that cover 10,000 square miles. The salinas are shallow, elliptical bays similar to those in Nebraska, Kansas and the Carolinas though oriented differently. These are all possibly cometary or icemass falls. Other sources state that the Odessa Meteorite craters were created fifty thousand years ago. Others say sixty-three thousand years ago. Geologic dating is marvelous.

Are we getting too many contemporary Meteor or Asteroid showers? The Heavens were certainly very active in this period. How many other meteor showers are dated from this time?

Remember that we have several dates for these events. Take your pick. Around 10,000 BC does seem to fit in quite well though.

There is massive evidence for meltwater flooding in the Gulf of California beyond the mouth of the Colorado River that must have come in a big surge down the Colorado River Valley.

The remains of many late Ice Age birds and animals have been unearthed from asphalt in McKittrick in Kern County in California from the end of the last Ice Age. What caused the massive stampeding across North America and apparently many other places on Earth? What were these thousands of animals stampeding from? Something that was horrendous coming from the skies or sweeping across the land such as tsunamias or mudslides?

Mastodon skeletons were discovered still standing upright and engulfed in great heaps of volcanic ash and sand from the end of the last Ice Age in the San Pedro Valley in Los Angeles County in California.

The United States geological survey found a bed of now petrified fish on a former sea bottom where more than one billion fish died within a four-mile area around 10,000 BC in the sea off Santa Barbara in Santa Barbara County in California. These fish averaged 6 to 8 inches in length and covered four square miles of bay bottom. How did they get trapped there? What force pushed these fish as well as seawater into this tiny bay? Was it a tsunami? We had animal stampedes in this period so did we also have fish stampedes?

Remains of typical Late Ice Age birds and animals have been unearthed from asphalt from the end of the last Ice Age in Carpinteria in Santa Barbara County in California.

The remains of dwarf mammoths have been found on Santa Ventura Island in the sea off Los Angeles in California. These dwarf mammoths were roasted in ancient pit fires. The mammoths were roasted and eaten by humanlike creatures that were reported to be giants with double rows of teeth the same as on Santa Rosa Island and Lompock Rancho nearby? We know that the dwarf mammoth has been extinct since the Cataclysm of eight to ten thousand BC. The double dentition has been reported elsewhere as well.

Along the ridge tops of the Santa Paula Mountains in Ventura County in California the remains of whales and other marine animals have been found in a totally confused state. At one site two thousand feet above sea level the remains of a large seal were found. Was it out for a walk? What sort of tides did this?

Valentine, 1969, mentions massive piles of mastodon and saber-toothed tiger bones that were discovered in Florida dating back twelve thousand years.

Elliptical craters all pointing in the one direction and the same as the Carolina Bays are found in Florida. Remember you have google to go searching with as well.

Marine deposits from the end of the last Ice Age occur at altitudes of 160 feet in what is now Georgia. Did fish take to walking up hill?

Geologists regard the Great Lakes of North America as having been formed at the end of the last Ice Age when the continental glaciers retreated and the depressions freed from the glaciers became lakes. In the last two hundred years Niagara Falls has retreated from Lake Ontario toward Lake Erie at the rate of five feet annually. If this process has been going on at the same rate since the last Glacial Period seven thousand years were needed to move Niagara Falls from the mouth of the gorge at Queenstown to its present position. With the advent of the last Glacial Period the Indians retreated southward returning to the north when the ice uncovered the ground and when the Great Lakes emerged, the basin of the St Lawrence was formed and Niagara began its retreat toward Lake Erie. Niagara Gorge was created in this period.

Sir Charles Lyell showed that the glaciers do not cut out holes like the depressions in which the Great Lakes lie. He shows that these lakes are not due to a sinking down of the crust of the Earth because the strata are continuous and unbroken beneath them. There is a continuous belt of such lakes reaching from the northwestern part of the United States through the Hudson Bay territory, Canada, and Maine to Finland. This belt does not reach below fifty degrees north latitude in Europe and forty degrees north in America. The depth of Lake Superior goes down to nine hundred feet deep and could not have been caused by digging out by an ice-sheet.

The encroachments of the waves upon the shores of the Great Lakes reveal whole forests of the buried trunks of the white cedar under the lake surfaces.

Richard Firestone, Allen West and Simon Warwick-Smith state that the Chippewa Basin at the deepest part of Lake Michigan is actually a meteor or bolide crater. It is 925 feet deep and 65 miles wide or 104 kilometres across and has a rough irregular bottom unlike the other lake bottoms as well as cracks and ridges which radiate from the centre. If a glacier had caused them they would be parallel. This would make this the fourth largest known impact crater on Earth. To some that is, we have several more in this listing. Other sources state that there are quite a few much larger craters. This Lake Michigan Crater was formed between eleven and twelve thousand years ago. Some say sixteen to thirteen thousand years ago. The eleven to twelve thousand year date ties in with other impact craters in the area as well as other impact produced phenomena that occurred in this period.

In the northeastern part of Lake Superior there is a submerged impact crater that is half a mile wide that dates from the end of the last Ice Age.

This is two impact craters in the Great Lakes region from this period which are the Chippewa Basin in Lake Michigan and the Lake Superior Crater. Are there others in this region? Three if you count the Charity Shoals Crater from the previous millennium if you believe the dating. How many others in fact?

Around Wichita in Kansas there are bays similar to the Nebraska and Carolina ones that are shallow, elliptical, flatbottomed and up to three miles long. They are oriented to the northeast the same as the Nebraska Bays.

Research by Richard Firestone, Allen West and Simon Warwick-Smith has indicated that one hundred and eleven stony meteorite finds in Kansas showed them forming three distinct streaks that were pointed towards Lake Michigan.

Out on the Great Plains of Nebraska there are baylike features that are called rainwater basins that are up to four miles long. There are hundreds of them and they are elliptical, shallow and flatbottomed with rims that overlap other bays. These bays, unlike the Carolina ones, point to the northeast rather than the northwest pointing towards the northern part of Lake Michigan.

There are numerous elliptical lakes in Bladen County and Columbus County in North Carolina. They all face from northwest to southeast and appear to overlay each other. There are great meteorite craters found in the Carolinas,

Georgia and Virginia dating from ten to twelve thousand years ago or 10,000 BC to 8,000 BC.

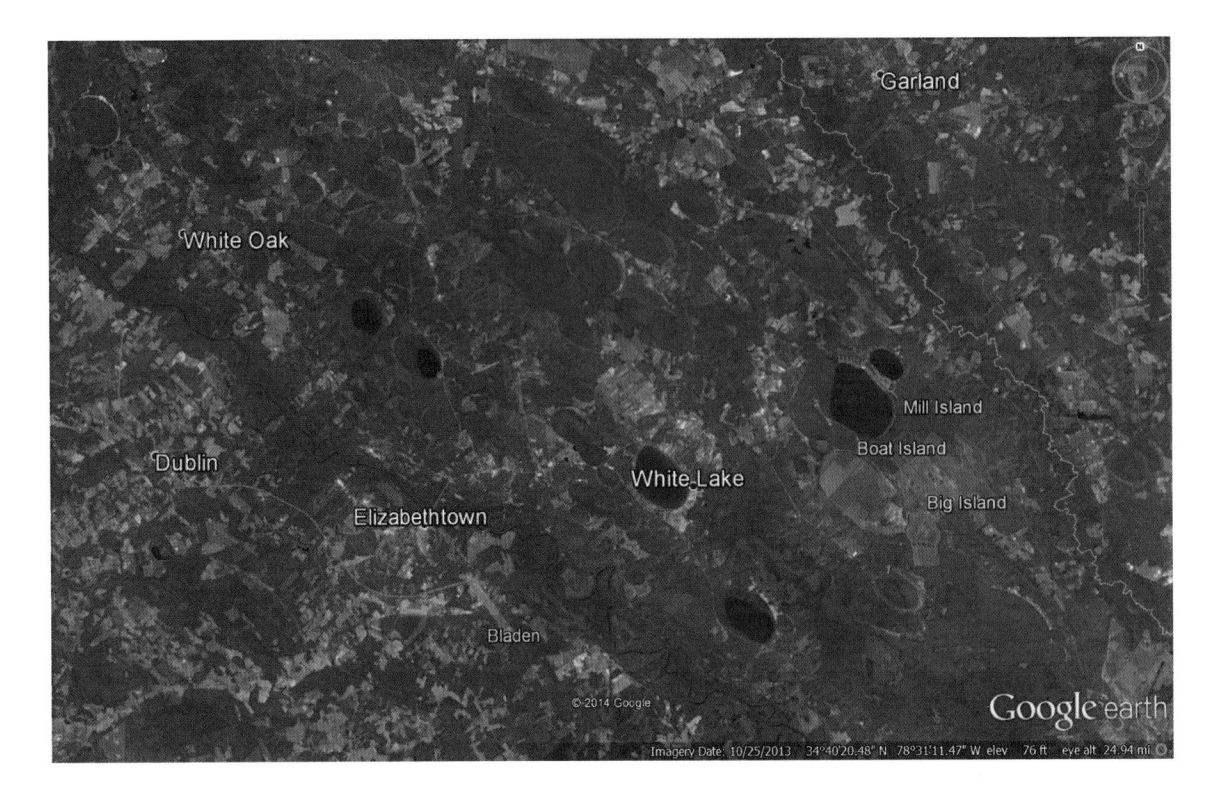

Carolina Bays in Bladen County/ Columbus County North Carolina. Notice how several of the dry lakes overlap which wind–caused lakes cannot do. These elliptical lake systems are world-wide.

    Just to refresh your memories I will repeat what I wrote about the Carolina Bays again. The Carolina Bays for example are elliptical and average half a mile in diameter with elevated rims and depressions 25 to 50 feet deep. The northwestern ends have low rims and the southeastern rims are higher as if something skidded at an angle in that direction. The Carolina Bays have also been called the Carolina Craterfield. In 1931 the authorities in the two adjacent states of North Carolina and South Carolina decide to try out a new photogrammetric survey and commissioned a company specializing in aerial photography to undertake the task. The stretch between Florida and Cape Hatteras and its hinterland were exposed to the aerial camera. When the films were enlarged and examined in the stereo-comparator they created a sensation. The pilots who saw the pictures were startlingly reminded of the battlefields of the 1914-1918 War. Huge circular or oval shapes sometimes overlapping that could be clearly identified as enlarged mud filled craters caused by the impact of gigantic boulders characterized all the pictures. The existence of the Bays had been completely unknown in the flat marshy land that was traversed by troughs or bays. The origin of the bays was now suspected. There were three thousand such troughs. Half of them are longer than 1,300 feet and more than one hundred are longer than 5,250 feet. They cover an elongated ellipse of 63,500 miles. Only a small portion of this is on land. The rest is on the fragmented coastal strip. All the longitudinal axes are parallel indicating that the fragments must have consisted of boulders traveling parallel to each other. Even though the craters are 11,000 years old most of the troughs still display remains of the thrust wall in the Southeast. The boulders must have come from the Northwest. The cause of the craters or bays was attributed to the Carolina Meteorites. Because the bays are elliptical as well as circular indicates that the descents and angles of impact must have differed considerably. Only a very large and disintegrating celestial body could have done this. A meteorite shower, if one were large enough would have produced identical holes, either circular or oval, not mixed. A descending and exploding large cosmic body could only produce the Carolina Bays. When the nucleus is traveling ahead of the tail of fragments the craters overlap. Another theory is that the bays were formed by masses of ice ejected from the area of the Great Lakes and that is why no meteoric masses have been found in them. In some of the bays on the low rims for the bays are shallow and elliptical generally with their end points parallel to each other, meteoric black glass fragments have been found that are the same as the meteoric glass fragments found in Clovis sites before their disappearence. The sand around the bay rims is white as well and this indicates massive exposure to heat as the sand was blasted away from the impact zone. This rules out ice masses from the Great Lakes and sounds more like cometary fragments. There are over one hundred thousand meteor craters concentrated along the Georgia, Florida, Virginia and North and South Carolina coasts but also extending North and South indicating a great meteor swarm once struck the area.

    Near Camden in Kershaw County in South Carolina at a depth of fourteen feet are great numbers of large fallen trees that all give the impression that their end was sudden and horrendous. It appears as if some massive force coming from the northwest pushed them over suddenly. Incidentally the Carolina Bays also indicate that the cometary bombardment came crashing down from the northwest! This is the same as the two great elliptical craters on the Puerto Rico Trench.

    Mr. J. Harlan Bretz uncovered superflood evidence from this period in a narrow channel near Portland in Oregon. Here glacial waters from glacial Lake Missoula had reached depths of 400 to 500 feet and possibly up to one thousand feet! The ice dam that had created this flow was estimated to be nearly half a mile high and when it burst the Missoula Superfloods were created. Other superfloods were occurring across North America as other ice dams were collapsing in all directions. At its largest Lake Missoula was four times larger than Lake Erie.

    In 10,000 BC someone left behind several pairs of excellently made woven sandals in a cave on Fort Rock in Lake County in Oregon. Fort Rock is so named because the rock formations on top of Fort Rock resemble a fort that has been subject to intense heat, melted and changed. The form of melting here is rare for it could only be caused by a tremendous heat source coming down from above. A huge thermal or nuclear explosion could do this. Or exploding cosmic matter hurtling down from the heavens? This is called rock vitrification and as we go on is not as rare as we originally thought. Some sources state that the site dates back to 12,000 BC.

In the ocean bed off Puerto Rico there are great meteorite craters dating from ten to twelve thousand years ago. Near the stump that is all that remains of the fractured Puerto Rico Plateau there are two great holes nearly 23,000 feet in depth. These holes are in the centre of the fragmented coastal area not very far from the southern edge of the submarine landmass that originally blocked the Gulf Stream before its submergence. The Puerto Rico Trench itself has a depth of 30,000 feet or 9,000 metres and encircles the southern part of the central area of the catastrophe. The two large holes are driven into the sima floor of the Atlantic Basin. The two holes are adjacent to each other and are similar in size and shape. Both are roughly oval and in both the major axes of the ellipses runs from northwest to southeast. This would suggest that the objects that struck like cosmic artillery shells and gouged out these deep sea holes came either from the southeast or northwest. Nothing of relevance is found in the southeast but in the northwest there is a subsided coastal strip. The stupendous force of this meteoric impact fragmenting the land so hard so that it sank beneath the sea. We may then deduce that the celestial body came from the northwest. The comet or asteroid appeared to have come from the northwest out of the evening sky and would have to have been traveling at a speed greater than that of the Earth in its orbit otherwise it would not have overtaken the planet.

On the Puerto Rico Plateau a million cubic miles of the tough viscous material of the sima cover has been displaced and fragmented. The land in the northwest remained. The cosmic bombardment that caused this must have left traces to the northwest as parts of it grazed the area. These could well be the Carolina Bays.

A mammoth staircase with steps two metres forty centimetres apart and descending eight kilometres into the deep sea and cut into the Continental shelf north of Puerto Rico was found. A dive made by the French submarine "Archimede" off the northern coast of Puerto Rico discovered the flight of steps cut into the steep sides of the continental shelf at greater depths than any previous finds. The steps continue for five miles under the sea and the steps are six and a half feet high each and only sixteen inches apart. Had this stair case been put in upside down in ancient times or was it tipped over by the massive impacts nearby?

A mastodon kill site near Sequim in Clallam County in Washington is at least twelve thousand years old.

At The Dalles or rapids of the St Croix River in Wisconsin there was an outburst of trap rock, which came up through open fissures breaking the continuity of the strata without tilting them into inclined planes. It would appear as if the Earth in the first place cracked into deep clefts and the igneous matter took advantage of these breaks to rise to the surface. It caught masses of the sandstone in its midst and hardened around them. These great clefts are lines radiating southwestwardly from Lake Superior as if it was the seat of the disturbance that created them. Were the Great Lakes created by a massive meteorite bombardment? When was this massive meteoric bombardment though? Could we well postulate the period of eight to ten thousand BC when the other geological events occurred in the same area? Otherwise we have pinpoint returning meteorite showers which is even more improbable.

At The Dalles or rapids of the St Croix River there was an outburst of trap rock, which came up through open fissures breaking the continuity of the strata without tilting them into inclined planes. It would appear as if the Earth in the first place cracked into deep clefts and the igneous matter took advantage of these breaks to rise to the surface. It caught masses of the sandstone in its midst and hardened around them. These great clefts are lines radiating southwestwardly from Lake Superior as if it was the seat of the disturbance that created them. Were the Great Lakes created by a massive meteorite bombardment? When was this massive meteoric bombardment though?

Could we well postulate the period of eight to ten thousand BC when the other geological events occurred in the same area?

In the ocean depths 200 miles east of Caracas in Venezuela probes have shown that the sediment on the sea bottom was exposed to air and sunlight twelve thousand years ago. This would seem to indicate that waves a mile deep raced across the Earth in that area to expose that part of the seabottom. Those waves could break against mountains at a height of six to seven thousand feet and would be sufficient enough to wipe out the human race in most parts of the world.

The oldest remains from Jericho in the West Bank show that Natufians occupied the area around 10,000 BC because it was a natural spring and oasis during the drought period of the Younger Dryas. Soon there was a large community here covering four hectares when the average for a community was less than one hectare. This was comprised of beehive houses densely packed together and separated by courtyards and narrow alleyways. This village huddled behind a massive stone wall with a masonry tower bordered by a rock-cut ditch nearly three metres deep and over three metres across. At Jericho there are circular ruins resembling from above interlocking wheels or gears. These are similar to other prehistoric ruins dating back to 10,000 years before the present day at Mnajdra in Malta and in the Canary Islands in the Atlantic Ocean. Pilots flying over areas where the Continental shelf is only one hundred feet off the surface have reported underwater ruins with similar outlines in other parts of the Atlantic Ocean. Stone city walls and towers were built at Jericho in 7,000 BC. Some reports state that there was definite proof of a large community here by 8,000 BC. By then massive stone walls thirteen feet thick and ten feet high were encircling an area of ten acres. There was an encircling stonewall with a stone tower thirty feet high. Inside the stone tower was an internal spiral staircase thirty feet of which was still standing ten thousand years later. Shallow underground dwellings were common. This is the same as at Maadi opposite Giza in Egypt. The first tower in the Mediterranean area was built at Jericho and is thirty feet high. There are also walls sixteen feet high that were replaced by walls 22 feet tall. Skilled engineers must have done this work yet they did not know pottery at all and ate off plates and dishes made of flint and used stone vessels where we use crockery. Their knives, scrapers, saws and augurs were made of flint or obsidian. The houses were shaped like halved eggs and were generally two storeys, the walls being of oval bricks and a floor of burnt stucco. The corners were rounded to avoid collecting dust. In 8,000 BC the inhabitants of Jericho were constructing enormous fortification walls, gouging out vast trenches in the hard bedrock and erecting a gigantic stone tower in defense against an unknown enemy. Engineering projects such as this require a great amount of social structure and coordination as well as social stability. Ten human skulls were found here. The features of the dead were modeled with traces of colour and with shells for eyes. These skulls were buried under the floors of the houses and had been made by modeling plaster over the skulls to produce lifelike effects.

9,880 BC. Allerod Oscillation.

The Allerod Oscillation occurred in Europe and North America. This was where in the period of 9,880 BC to 8,850 BC there was a temporary increase in warmth causing glaciation to retreat allowing forest trees to establish themselves in ice-free zones. This prevented minor glaciation from building up into major glaciation due to the increasing of the forest area, and the subsequent rise of general temperatures and humidity levels which would have become self perpetuating. The Ice Age might still be here if it weren't for the Allerod Oscillation and the impactoids that crashed to Earth in this period.

During the Allerod Oscillation sea temperatures dropped in the North Atlantic, in the western North Pacific, in the South China Sea and even the tropical Sulu Sea between the Philippines and Borneo.

After this last advance of the glaciers from the polar cap the climate became warmer and by 8,000 BC, the Mesolithic age, the ice sheet retreated and opened the way to new lands for men, animals and plants. The climates assumed their present characteristics around 10,000 BC to 8,000 BC. Europe and North America got considerably warmer than they are now.

All over Europe during the latter part of the Ice Age from 12,000 BC to 8,000 BC great rivers swollen from the melting ice and rain carried large quantities of gravel and silt down from the glaciers and icecaps. These overloaded rivers regularly silted up, flooded and changed their courses. Over the years they filled in the valleys with many yards of debris creating wide deluge plains.

Mud from Lake Knockacran in Ireland traced to the last ice sheet was found to be 11,787 years old around 9,787 BC.

The End of the Younger Dryas.

Recently obtained ice cores from Greenland indicate that the Younger Dryas ended in only two to three years during a rapid retreat of sea ice in the northern Atlantic Ocean and a fifty percent increase in rainfall and snowfall. Some unknown trigger in the North Atlantic raised sea temperatures rapidly. Once again we have to average out the dates and allow for longer occurrence spans or greater flexibility in dating periods. This was around 9,700 BC.

A scientific association meeting in Vienna calculated that in 9,684 BC, our Earth, spinning at the Poles, sustained a collision with the head of a tremendous comet. The terrible impact caused the Earth to lurch and tilt violently at the axis. There followed a revolutionary change in world climate from lush, steamy warmth of tropical forest or marsh to the intense cold of the Ice Ages with glaciers covering the North and farther Southern Hemispheres. This period around 9,684 BC incidentally is when Plato wrote that the fabled continent of Atlantis was destroyed in one day and one night. Plato stated that Atlantis disappeared in 9,600 BC. Is there any connection? Was it a direct hit? What is 84 years?

Please acknowledge that all of the impacts that I have dated to ten thousand BC might well be at the same time as this one as ten thousand BC covers the period up to nine thousand BC. It might even cover the period to 8,000 BC.

The renowned Geologist Professor Alexander Tollman of the University of Vienna studied the worldwide distribution of tektites, splinters of molten rock thrown up by an impact to the Earth's surface. He was convinced that the planet must have suffered a major space collision around 10,000 BC to 9,000 BC. A sudden increase in radioactive Carbon-14 found in fossilized trees from the period backs up his conclusions. Tollman believes that seven pieces of a disintegrating comet intersected with Earth's orbit with calamitous results. This impact caused earthquakes, geological deformation, a vapor plume and tidal waves. It also caused the extinction of the mammoth and the saber-toothed tiger. The Professor suggested 9,600 BC as the date of the event.

Professor Emiliani identified a rapid melting of glaciers at about 9,600 BC as the cause of flooding for ten years.

Laboratory analysis of two deep sea cores collected in the Gulf of Mexico by University of Miami scientists proves that the entire water level of the World substantially rose in the vicinity of eleven thousand six hundred years ago, 9,600 BC. Further developments in research in the Gulf of Mexico show from the study of fossil shells of *Foraminifera*, a marine life form, that between eleven to twelve thousand years ago there was a substantial increase of twenty per cent in the warmth and a twenty per cent decrease in the salinity of the water. This indicates that there was a profound change in the levels, salinity and the temperatures of the oceans as glaciers started to melt at this time. But doesn't this conflict with other reports that the Gulf of Mexico was a large plain? Maybe part of it was. Parts of the Gulf of Mexico go down to 14,000 feet. Remember that the dates for extinctions conflict in this period they are so close together. 9,600 BC could be easily 8,600 BC.

On the island of Spitzbergen, Svalbard, palm leaves ten feet and twelve feet long have been fossilized along with fossilized tropical marine crustaceans. This is the climate of the Bay of Bengal or the Caribbean, not the frozen areas of eighty degrees north where ships can only reach Spitzbergen through the ice for only two or three months of the year.

In Spitzbergen at 78 degrees 56 North there were found the remains of similar plants to beeches, oaks, poplars, maples, walnuts, magnolias, limes and vines. Interspersed within the muck depths and sometimes through the very piles of bones and tusks themselves are layers of volcanic ash. This indicates volcanic eruptions of tremendous proportions.

Whole islands just north of Siberia are formed from the bones of Pleistocene animals apparently swept northward from the continent.

In northeastern Siberia, above the Polar Circle, above latitude 75 degrees, there was no perennial ice during the last Ice Age.

Siberia during the last Ice Age was warmer than it is today. Of thirty-eight species known to have lived in Siberia before 9,600 BC including mammoths, giant deer, cave lions and cave hyenas thirty-four were adapted to temperate regions. This indicates that Siberia was temperate and much warmer that it is now. Siberia is now a sub-temperate region of frozen tundra, windblown steppes and at lower latitudes dense coniferous forests. Of the thirty four animals species living in Siberia prior to the catastrophes of the eleventh Millennium BC, including Ossip's Mammoth, giant deer, cave hyena and cave lions, no less than 28 were adapted only to temperate conditions. The farther north one goes the more mammoth and other remains increase in number. Some of the New Siberian Islands were described by explorers as made up of almost entirely of mammoth bones and tusks. The nineteenth century French zoologist

Georges Cuvier stated that the eternal frost did not previously exist in those parts in which the animals were frozen for they could not have survived in such temperatures. The same instant that these creatures became bereft of life was the moment that they became frozen. And these poor creatures were clambering onto the peaks of hills that became islands as the weather suddenly changed along with the sudden rise in water levels. Or they were being swept into gigantic mounds or islands of their own remains as well as every other living thing around them.

The northern Siberian plains supported vast numbers of rhinoceroses, antelopes, horses, bison and other herbivorous creatures while a variety of carnivores including the saber-toothed tiger preyed on them. Like the mammoth these animals ranged to the extreme north of Siberia, to the shores of the Arctic Ocean and yet further north to the Lyakhov and New Siberian islands, only a very short distance from the North Pole. Remember though that the North Pole was not in that position then. It was still in Hudson Bay.

Mammoth and rhinoceroses and other animals that are not normally associated with the climatic conditions of present day Siberia were entrapped in floods of freezing mud and preserved so quickly that undigested foods of plants no longer native to Siberia were found in their mouths and stomachs. This happened 11,000 to 10,000 years ago. Parts of Northern Siberia, Canada and Alaska are so covered with bones that some islands or highpoints where they went for refuge seem to be made entirely of their bones.

The Arctic explorer Baron Eduard von Toll found the remains of a saber-toothed tiger and a fruit tree that had been ninety feet tall when standing on one of the New Siberian Islands. The tree was well preserved in the permafrost with its roots and seeds. Green leaves and ripe fruits still clung to its branches. The only representative of tree vegetation on the island today is a willow that grows one inch tall. How do you snap freeze a tree? You move it suddenly to a new climatic zone.

Northeast Siberia was not covered in ice in the last Ice Age. The climate has only changed drastically since the last Ice Age and animals that once lived in this region do not live there now. Plants grew there that cannot grow there now. This must have occurred suddenly as all of the mammoths of Siberia perished. The mammoths belonged to the family of elephants. The tusks were up to ten feet long. Their teeth were highly evolved and their density was greater than in any other stage in the evolution of the elephant. They were virtually a super elephant.

The Siberian mammoth must have perished quickly as they were frozen without putrefying. Their skin, hair and flesh are exceptionally well preserved. They appeared to have been frozen as soon as they were killed. They could not have lived at the temperatures of a place of eternal frost. It was at one and the same moment that the country became covered with ice and the animals perished. It was instantaneous with no gradiation.

The mammoths did not die of starvation either as the remains of undigested grass and leaves were found between their teeth and in their stomachs. Many of the leaves and twigs found in their stomachs do not grow in the regions where the mammoths died but far to the south, over one thousand miles away. It appears that the climate has changed radically since the death of the mammoths. They were found encased in blocks of ice. The change of temperature must have followed their deaths very closely or in fact caused them. After storms in the Arctic tusks of mammoths are washed up on the shores of Arctic islands thus proving that a part of the land where the mammoths once lived is now drowned and covered by the Arctic Ocean. The mammoths died suddenly, in intense cold and in great numbers. Death came so quick to the mammoths that swallowed vegetation is undigested. The mammoths died in midswallow. Grasses, bluebells, buttercups, tender sedges and wild beans have been found undeteriorated in their mouths and stomachs. Such flora does not grow in Siberia today and its presence there at the end of the eleventh Millennium indicates that the region had a pleasant and productive climate, ranging from temperate to even warm. As the last Ice Age ended in other parts of the World it suddenly started in Siberia. If they had not been frozen as soon as they were killed putrefaction would have decomposed them. This eternal frost though would not have prevailed where they had died for they could not have lived in such a temperature. Nor with frozen ground would they be able to sink into it. It was therefore at the same instant when these animals perished that the country they inhabited was rendered Glacial. These events must have been sudden, instantaneous and without any graduation.

Today the Siberian tundra is a desolate expanse with a winter temperature lower than the North Pole. The annual mean is 16 degrees below zero and a lower limit of 49 degrees centigrade in January. The mammoth could not have lived in such a hostile and cold environment for examination of its remains show that contrary to what many people believe it was accustomed to living in a temperate zone like the horse, the tiger, the bison, the antelope and the other mammals which were involved in the same general destruction. Examination of the foodstuffs found in their stomachs makes it clear that Siberia in their day was a mild region of luxuriant vegetation.

Admiral Wrangel tells that the remains of elephants, rhinoceroses etc are heaped up so much in certain parts of Siberia that he and his men climbed over ridges and mounds composed entirely of their bones.

The Siberian mammoth died simultaneously with the European and Alaskan mammoth. A number of authorities attribute the violent intermingling of mammoth carcasses and broken trees in Siberia to great tidal waves that uprooted forests and buried the tangled carnage in a deluge of mud that quickly froze. In the Polar region this deluge froze solid and has preserved the evidence in permafrost to this day.

Melting ice sheets at the end of the last Ice Age would have caused sea levels to suddenly rise by about three hundred feet.

The glaciers of the Ice Age covered the greater part of North America and Europe while the north of Asia remained free. In America the plateau of ice stretched from latitude 40 degrees and even passed this line. In Europe it reached latitude 50 degrees. In northeastern Siberia, above the Polar Circle, above latitude 75 degrees, there was no perennial ice. Move the North Pole to the position of Hudson Bay and this all makes sense.

Between eleven thousand and nine thousand BC there were numerous upheavals in the northern regions of Siberia and Alaska around the edge of the Arctic Circle. Uncountable numbers of large animals have been found; many carcasses still intact as well as astonishing quantities of perfectly preserved Mammoth tusks. Hundreds of thousands of individual creatures must have frozen immediately after death and the remains frozen otherwise the meat and ivory would have spoiled. The mammoth meat appears so fresh that it has apparently been offered in restaurants in Fairbanks, Alaska, and has been used to feed sled Dogs.

It is estimated that some ten million animals lay buried along the rivers of northern Siberia. Ten million very large animals that ate half a ton of feed per day were now dead. Try and get that much feed out of moss and snow. The stomach contents found were of small branches and leaves and not of tundra plants.

It seems that the Ice Age had just appeared in Siberia when it had left everywhere else. This was due to a sudden earth tilt where the land masses were moved into different climatic zones.

D. S. Allen and J. B. Delair proposed that a massive interplanetary visitor that they named Phaeton passed close to the Earth and Mars around 9,500 BC. Phaeton was created in an astronomically near supernova event and was a portion of exploded star matter.

D.S Allen and J.B. Delair also stated that the earth's axis was pulled into a tilt by fragments of a supernova in the Vela star system that had been blasted into our solar system in 9,500 BC. This was when the Pleistocene period ended and the Holocene period began.

Some state that the Younger Dryas, the Mini Ice Age, ended around 9,500 BC. Was this period a cosmic free for all? Or was there a free for all of explanations for the unknown events that were striking the earth.

In 1975 marine scientists from the University of Miami in Florida discovered traces of fossils and limestones taken right off the surface of the Atlantic seafloor that showed traces of rainwater. This can only occur on the surface of the sea, not thousands of feet down under the ocean. These scientists believe that a sudden flood of icy waters caused deepsea marine life to drastically alter their characteristics because of some event around 9,500 BC.

In a matter of only a few days the planet had altered its orbit and completely changed the climate of all of the landmasses as they were shifted relative to where the Polar weather masses would be.

Nothing would ever be found of the continent often referred to as Atlantis on the Mid Atlantic Ridge as it had been effectively vapourized and the remnants apart from a few mountain peaks submerged.

Strong winds and storms as well as rapidly changing weather patterns spread death across the Americas causing the extinction of the American horse, the American elephant, the American camel, the mammoth and mastodon and the five toed llama amongst many others. The age of the giant mammals was effectively over. This same death and destruction occurred in Europe and Asia and even Australia.

Luna Jheel or Luna Zeel, Zeel is Urdu for lake, is an impact crater dating back eleven thousand five hundred years from 9,500 BC. Luna Jheel is 2.15 kilometres in diameter. Shocked quartz, impact glass and other signs of impacts have been found here. Luna Lake or Luna Zeel is at Luna near Bhuj on the Banni Plains on the Rann of Kutch in Gujarat in India.

Shanidar is a huge cave situated high above the Greater Zab River about 520 kilometres northeast of Abu Hureyra in the mountains of Kurdistan in Iraq. Several skeletons of *Neanderthals* were found who had died from a roof fall and had been buried ritualistically. Ashes and food remains over the graves hinted at a funeral feast and as well there were eight different types of pollen from wildflowers that suggested that the flowers covered the dead. Two of the skeletons were of an old man and a disabled woman indicating that the community looked after its frail and aged as well as having a possible religious belief in the afterlife. Ralph Solecki also discovered a slim almond shaped piece of copper with two equally spaced perforations at its end so that it could be worn as a pendant around the neck. The stratum it was found in was from around 9,500 BC.

The Ouro Ndia impact crater in Mali in Africa is eleven thousand five hundred years old and three kilometres in diameter.

The Gogui Crater in Mauritania in northern Africa is eleven thousand five hundred years old and six hundred metres in diameter.

The temperate climate of Siberia was snap frozen as the Arctic weather patterns suddenly moved to their new positions as the Earth moved. The same weather disturbance happened to the Poles which were previously semitropical in places as has been discovered by fossil evidence. Isn't this the same as the report for 9,600 BC? We are getting conflicting dates again but they are close enough to be the same date?

At Shakta Cave in the Pamirs or Pamyrs of Siberia at an altitude of 14,000 feet the Russian Archaeologist V. A. Ranov found drawings made with red mineral paint depicting a bear, boar and ostrich, none of which can survive in the present Arctic temperatures. A clue to the age of the drawings, the highest altitude prehistoric drawings in the World, was found at Markansu where settlers left artifacts and ash. The latter were of burnt cedar and birch, which do not grow in the region today either. Carbon-14 dating showed 9,500 years. The Pamir Mountain range was raised from being a plain and the remains of reindeer, plains dwelling animals, were found at heights that reindeer could never have lived at. Then the area was suddenly frozen. Here was the climate moving again.

The Zhongcangxiang Crater in Tibet is eleven thousand five hundred years old and is five hundred metres in diameter.

In Turkey near the borders of Syria and Iraq is the site of Gobekli Tepe, "The Hill of the Navel". Three square buildings were constructed that are eight to twenty metres in size, one behind the other close to the summit of a south-facing hill. The structures were decorated with lines and rings of T-shaped pillars around three metres tall. On their surfaces were sculpted high reliefs of foxes, cranes, waterfowl, spiders, lizards, crocodiles, serpents and birds described as raptors. There is also a large cat whose elongated form is carved to flow with the curved shape of the megalith. Other pillars appear like human forms. The T-shaped pillars only weigh a few tons each but there is a nine metre tall one still in the nearby quarry that weighs fifty tons. There are no habitations at the site and it seems to be purely a ritual site. Excavations here are revealing new ruins covering quite an area. All of these ruins were intentionally buried around 9,500 BC. Was this to protect them?

There are craterlike shallow depressions from the period of 9,500 BC in the area of Point Barrow in Alaska. These are the same type of craters found on the Atlantic coastal plain and the Carolina Bays. There are similar shallow craters in Bolivia and Holland from the same period. These shallow craters in some cases are regarded as being caused by ice masses hitting the Earth such as found in comets. These are oriented shallow depressions or lakes in the permafrost near Point Barrow that are aligned northwest to southwest. These are the same age and the same direction and shape as the Carolina Bays and are from the same period. Interestingly enough there are the ruins of an ancient city here above the present Arctic Circle. Are our dates correct for these ruins or are they much older than we know? Yes, we have found this report before with a different date. That is geology for you.

The Crestone crater near Hartsel in Saguache County in Colorado is eleven thousand five hundred years old and is 108 metres in diameter.

So the known tally of 9,500 BC is Luna Jheel Crater in India, Ouro Ndia impact crater in Mali, the Gogui Crater in Mauritania, the Zhongcangxiang Crater in Tibet, the Point Barrow Craters in Alaska and the Crestone Crater in Colorado. Does this seem to be a normal year?

The Younger Dryas suddenly ended in 9,400 BC, and 9,300 BC up to 9,700 BC, when there was a sudden change from dry to wet climate in the Near

East along with the elevation of local water tables. Which one is true? How many endings were there to the Younger Dryas?

Everyone now has a theory and these are all different theories that I am presenting along with the evidence.

Around 9,400 BC the Gulf Stream in the Atlantic Ocean suddenly restarted itself thereby increasing the temperatures of Western Europe and altering world climate patterns again. This was the official end of the Younger Dryas mini-Ice Age and the return to warmer temperatures. This warming occurred in only fifty years and melted the ice over the now rapidly saltifying freshwater that had been deposited into the North Atlantic as Lake Agassiz suddenly ceased to exist at the beginning of the Younger Dryas and had flooded into the Atlantic Ocean via the St Lawrence Valley.

Dr Rene Malaise believes that there was a ridge that had been a barrier to the Gulf Stream that landlocked the Arctic Ocean from Europe to Greenland. When the land barrier sank the Gulf Stream reached the Arctic Ocean after travelling along the coast of Europe and the British Isles and Ireland and the Ice Age ended. Dr Malaise and the French Geologist J. Bourcart pointed out that two soil samples found in waters east and west of the northern chain were of a different nature. Mud from the eastern side was oceanic and mud from the western side was glacial in nature. This indicates that as the chain emerged from the water during the period of ice extension and marked out its limits between the warm Gulf Stream which was moving from the south along the western slopes, and the eastern current carrying icebergs along the eastern side of the Atlantic.

The Soviet scientist N. Zirov also postulated that the Atlantic landmass barred the warm Gulf Stream from the coasts of Europe, which were then covered in ice. Samples drawn from the Mid-Atlantic Ridge on the western side show ordinary oceanic mud and those on the eastern side revealed glacial origin showing that they were carried by icebergs. This was the same as Malaise and Bourcourt's findings.

Some geologists postulate that the Gulf Stream between Bimini and Florida was a fault resulting from an ancient earthquake.

When the lost landmass still existed in the North Atlantic and the Azores Plateau it blocked the way to the European coasts of the warm marine currents. When the landmass suddenly sunk the Gulf Stream gradually caused the ice to disappear by heating up the ocean as warm tropical waters moved northwards.

Professor Hull also stated that this Antillean land mass shut off the Gulf Stream from the Caribbean Sea and the Gulf of Mexico.

Ignatius Donnelly published in 1882 that with the removal of an Atlantean continental site the warm waters of what is now the Gulf Stream would be able to flow up to where they now are. The shores of Western Europe would no longer be intensely cold and the warmer waters of the tropics arriving in the north would have ended the Ice Age.

The second great meltwater spike began in 9,400 BC and never reached the Euxene Lake, now the Black Sea. The date of the second great meltwater spurt was corroborated by coral growth near Barbados. This time when the ice fields over Britain, Scandinavia, Holland, Northern Germany, Poland and Russia started to recede the "great lakes" that had filled with water after the 12,500 BC ice cap retreated, that were north of the Euxene Lake, were now trapped there as they were trapped in the depressions made by the weight of the ice. The peripheral bulge then directed the meltwaters away from the upper reaches of the Dniester, Dnieper, Don and Volga Rivers to flow westwards away from the Black Sea towards Poland and over where Berlin in Germany is now to the North Sea. This sideways flow of the meltwater discharge prevented any more of it to flow into the Black Sea which triggered the dessication of the Euxene Lake as it commenced to dry up. As the surface of the Euxene Lake dropped the corresponding sea levels of the world's oceans rose dramatically from the meltwater. Eventually by 5,600 BC the shoreline of the Euxene Lake was 350 feet below the level of the Mediterranean Sea behind its Bosporus dam. This would be when the Euxene Lake would become the Black Sea.

At the end of the Younger Dryas warmer temperatures and increased rainfall returned to the eastern Mediterranean. The dry cold northeasterly winds were replaced by moist westerlies activated by the reactivated Gulf Stream. Forests returned to Anatolia, Syria, Iran, Iraq and the Jordan Valley. The main trees were oaks and pistachios.

At the end of the Younger Dryas the northern Sahara became dryer as the jet stream moved northwards and accentuated the dryness there. Over the rest of the Sahara the rains improved and between 8,000 BC and 5,500 BC lakes throughout East Africa and the Sahara increased massively. Rainfall in these areas increased between 150 and 400 millimetres annually. This was created by a stronger Asian monsoon cycle.

Shortly before the end of the Younger Dryas a community was built on the north bank of the Melendez River at the foot of the volcano named Hasan Dag which eventually buried it. Asikli had been flooded by the rising water table at the end of the Younger Dryas. Evidence of advanced urban planning was evident in the layout of the houses, court yards and roadways with masonry consisting of both mud bricks and meticulously hewn blocks of volcanic tuff. The floors of the rooms were plastered and the checkered imprints of mats were impressed in them. There were four hundred houses in this community. There were also open air shops in which workmen knapped blades, scrapers and knives from the volcanic obsidian and in the scrap piles were fragments of polished mirrors. Asikli was buried by a sudden eruption of Hasan Dag that buried the village and its tilled fields indicating that agriculture was already in evidence at the end of the Younger Dryas.

A fossilized skull found at Punin at Chimborazo in Ecuador could be as old as that of *Tepexpan Man* of 9,300 BC. It is a longheaded mongoloid type of head that is common to Amerindians. Tepexpan Man was found in Mexico and is from the same period. The head is of a woman of Australoid type with a dolichocephalous skull which is a long skull. The skull was found alongside the remains of mastodons, camels and an extinct species of horse in beds of volcanic ash.

Tepexpan Man was discovered in Mexico State in Mexico. Helmut de Terra discovered the fossilized remains of early man along with the remains of an elephant from 9,300 BC. The skull was found in an alluvial flood plain from 10,000 to 15,000 years ago. Tepexpan man had a sloping forehead, flattish nose and wide zygomatic arches or cheekbones and was a typical Amerindian. Another report states that it was the body of a woman that was buried with that of an elephant and that they were both buried at the same time.

A longheaded mongoloid type of fossilized human skull was found at Pelican Lake in Alger County in Michigan that was probably as old as Tepexpan man from 9,300 BC. Who are these three people wandering across North, Central and South America? Two were dolichocephalic and one wasn't.

The Mullsjon Crater is a lake near Hjo in Vastra Gotaland County in Sweden. The crater is 5.3 kilometres in diameter and eleven thousand one hundred years old.

What then is the meteoric impact score for the Tenth Millenium? We have major impacts in the Bahamas, the Bermuda Impact, Beni and Araona in Bolivia, Amundsen Gulf, Baffin Bay, Bloody Creek Crater in Newfoundland, Hudson Bay, Lake Saimaa in southern Finland, Lake Racze on Wolin Island in Poland, meteorite craters found in the Carolinas, Georgia, Florida and Virginia, Sithylemenkat Crater near Bettles in Alaska, Barringer Meteor Crater in Arizona, Chippewa Basin at the deepest part of Lake Michigan, Lake Superior, Nebraska Bays, Carolina Bays, Puerto Rico, Odessa Craters in Texas, Zhongcangxiang Crater in Tibet, Luna Zeel in India, Ouro Ndia in Mali, Gogui Crater in Mauritania, Zhongcangxiang Crater in Tibet, Point Barrow in Alaska, Crestone crater in Colorado, Mullsjon Crater in Sweden. And these are all craters over half a kilometer in diameter. How many smaller ones were there?

Was this an unusual year? Or were there three unusual milleniums? Were these three milleniums actually combined in what would be one millennium?

9th Millenium BC.

Clube and Napier theorize that a giant comet settled into an earth-crossing orbit 50,000 years ago and for 30,000 years remained intact. 20,000 years ago, around 18,000 BC, a massive fragmentation event occurred and from 17,000 years ago, 15,000 BC, multimegaton fragments may have periodically collided with the Earth thus causing a reduction in glaciation. There were two large impacts in the eleventh Millennium BC, 11,000 BC, and the ninth

Millennium BC, 9,000 BC, which raised global temperatures so much that the Ice Age was brought decisively to an end.

Doctor William Stokes of the University of Utah states that there were anomalies in glacial movements at the end of the last Ice Age. The glaciers moved in directions opposite to what would be expected of them. In South Africa they moved north-south away from the equator, in Central Africa and Madagascar the ice moved northwards well within the Tropic zone, in northern India the ice moved northwards and in Australia and Tasmania the ice moved from south to north whereas in Brazil and Argentina it was towards the west. The glacial sheet in the Southern Hemisphere moved from the tropical regions of Africa towards the South Polar Region. In the Northern Hemisphere, the ice in India moved from the Equator towards the Himalayan Mountains and the higher latitudes.

Something very odd had happened to the South Pole and only comparatively recently as far as geology is concerned. Here we have had very good evidence of a planet-tilt and resultant climatic change.

The Hindu Kush Range on the eastern border of Afghanistan and India/Pakistan was lifted by a giant crustal displacement. The mountains suddenly rose several thousand feet according to some sources. This seemed to be happening in the previous millennium if you can believe the dates. Remember that I am only presenting data from numerous researchers. You can believe them or not.

In 1926 Claude Roux argued that in post-Glacial times most of North Africa was under the sea and the mountains of Morocco and Algeria constituted a peninsula. Eventually the land rose or the sea fell and the seas and lagoons dried up leaving the present day deserts and salt marshes.

In the post-glacial Quaternary period according to Claude Roux North Africa was a fertile peninsula that was bounded by great shallow lagoons stretching from the Atlantic and the Mediterranean to the southern Atlas Mountains. This mountainous stretch of land was very fertile and thickly populated with animal and plant life abounding. Eventually the lagoons receded towards the coast leaving lakes and salt marshes and the onslaught of the desertification. The lakes and salt marshes are the Schotts and Sebkas.

An ancient tradition from Oued Bou Merzoug in Constantine Province in Algeria stated that long ago a wicked people lived there in the Atlas Mountains and for their sins stones rained upon them from heaven. Memories of meteorite showers? Quite common mythologies?

Numerous remains of Pleistocene Era animals have been found in Algeria including mammoth and hippopotami that died around 9,000 BC. So mammoths and hippopotami also lived in the Tropics as well?

Professor Gregory stated that the Atlas Mountain Range in North Africa and the Alpine Mountain Range in Europe are fragments of an Atlantic continent which produced them.

Around 11,000 BC there was a large landmass in the North Atlantic Ocean called Appalachia by modern geologists. This large land mass connected Europe and North America via Greenland and Iceland. Sand, mud and gravel washed out to sea by its rivers and ground up on the shore built up a lot of half of the North American states. Some sediments were up to a mile thick but the entire land mass was now two miles, 3.2 kilometres, below the Atlantic Ocean. Could this have been Atlantis? Or another Neolithic lost civilization? Professor Geikie stated that biological evidence showed that Appalachia was still above sea level in post-glacial times. There was a colossal amount of sea bed scour between Greenland and North America in the area of Appalachia. This occurred at the same time as the entire North Atlantic Ocean floor subsided suddenly to a depth of 9,000 feet, 2,750 metres, between Greenland and Norway.

The Morro de Cuero Crater in Mendoza Province in Argentina is eleven thousand years old and six hundred metres in diameter.

Professor Cesare Emiliani a marine Geologist at the University of Miami in Florida stated that there had been a dramatic rise in water levels about 11,000 years ago, 9,000 BC, in the Atlantic Ocean.

During the late Nineteenth Century the United States ship "Dolphin", the German frigate "Gazelle" and the British ships "Hydra", "Porcupine" and "Challenger" mapped out the bottom of the Atlantic Ocean. The result is the revelation of a great elevation reaching from a point on the coast of the British islands southward to the coast of South America at Cape Orange, then southeast to the coast of Africa and then south to Tristan da Cunha. The elevation rises nine thousand feet higher above the Atlantic depths around it and in the Azores, St. Paul's Rocks, Ascension and Tristan da Cunha it reaches the surface of the water. The inequalities, the mountains and valleys, could never have been produced in accordance with the laws of the deposition of sediment nor by submarine elevation but must have been carved out by agencies working above water level. The officers of the "Challenger" found the entire ridge covered with volcanic deposits. The "Challenger" found that the inequalities, the mountains and valleys, could only have been produced while still above water and could not have been produced by submarine agencies such as deposit of matter only.

The German Atlantic expedition of 1935 revealed that the huge basin of the Atlantic is divided into two parts by a very long and massive submarine ridge extending from Iceland in the north to the Antarctic shelf in the south. The eastern basin has an average depth of 15,000 feet (4500 metres) and the western 21,000 feet (6500 metres). Depths along the ridge also go down to 23,000 feet (7,000 metres). The ridge between the two basins was called the "Dolphin's Back" or "Dolphin's Ridge" but is also known as the Atlantic Ridge or the Atlantic Plateau. The ridge rises on average 9,000 feet from the seabed. At 30 degrees west and 40 degrees north the ridge widens out into a huge sunken landmass like a great submarine mountain. Out of this mass rise sharp peaks that break the surface and appear as land. These peaks are the Azores Islands and some such as Pico Alto rise to 20,000 feet from the seabottom. This part is called the Azores Plateau. This is the region where the Gulf Stream crosses the Atlantic Ridge. The "Dolphin's Ridge" that the "Challenger" found was connected with the shore of South America north of the mouth of the Amazon.

The Gulf Stream would not move eastwards at all if the landmass were higher and above sea level. Europe would still be under glaciation as it was during the last Ice Age.

The Commander of the sloop "Gettysburg" prior to 1877 found that when about 150 miles from the Straits of Gibraltar the soundings decreased suddenly from 2,700 fathoms to 1,600 fathoms in the course of a few miles. Commander Gorringe believed that from his bank of soundings there was a submerged ridge or plateau connecting the island of Madeira and the Portuguese coast. Soundings were made five miles apart.

Seabed cores taken from the Mid-Atlantic Ridge in 1957 brought up the remains of freshwater plants from a depth of two miles. These freshwater plants had dropped two miles from being above sea level.

In one of the deep valleys of the Mid-Atlantic ridge named the Romanche Trench sands have been found that appear to have been formed by weathering when that part of the ridge was above sea-level.

In 1900 AD the research vessel "Gauss" hauled up a sediment core from a depth of 24,500 feet in depth from the Romanche Trench. The core measured one foot six inches. The core contained five strata. At the top was red clay, followed by three chalk-free continental sedimentary strata and lastly at the bottom globigerina ooze. This ooze is normally found at depths of 2,000 to 4,500 metres and cannot have been deeper than 4,500 metres when deposited. The seabed here must have subsided 2,800 metres or 9,400 feet. The Swedish ship the "Albatross" taking cores from the depth of two miles in the Romanche Trench fracture zone in the vicinity of the St. Peter and St. Paul Rocks brought up shallow water micro-organisms, preserved in bottom mud along with twigs, plants and tree bark all of which had rapidly descended into the depths.

St. Paul's Rocks in the Atlantic Ocean are reefs of ancient continental rocks that once belonged to a land that has foundered beneath the Sea.

In February 1969 where the Caribbean meets the Atlantic Ocean on the Aves Ridge, which stretches underwater from the Virgin Islands to Venezuela, more than one ton of granite was dredged up. Granite had never been dredged up from the open ocean before. The consensus of the Lamont-Doherty Geological Observatory team from Columbia University was that a lost continent had once risen from these waters or that a new continental mass was being built in the area. The granitic rocks were brought up from fifty sites along the Aves Ridge, an underwater ridge running from Venezuela to the Virgin Islands. Acid igneous rocks were found which are only found on continents.

Professor Gregory has reasoned that the islands of the Atlantic are irregularly arranged with the exception of those on the Faeroes Ridge, which forms the northern boundary of the Atlantic and the islands of the West Indies. The sinuous shape of the Atlantic moreover is not related to the mountains or the geographical grain of the adjacent lands. It cuts abruptly across the old mountain line, which runs across southern Ireland, southern England, and northern France and across the Cantabrian Mountains, which reach the Atlantic at the northwestern corner of Spain. There can be no doubt that these mountains must have originally extended into the Atlantic. On the coast of North America the Atlantic cuts with equal abruptness across the remains of mountain lines in Nova Scotia and Newfoundland, which must have extended eastward in the ocean. The floundering of a plateau land that must have once connected South America and Africa must have caused the intermediate basin of the Atlantic.

In 1969 a Duke University research expedition dredged 50 sites along an underwater ridge running from Venezuela to the Virgin Islands. They brought up granitic rocks that are normally found only on continents. Dr Bruce Heezen of the Lamont Geological Observatory stated that geologists normally believe that light granitic or acid igneous rocks are confined to the continents and that the crust of the Earth beneath the sea is primarily heavier dark coloured basaltic rock. The occurrence of the light-colored granitic rocks may support an old theory that a continent may have formerly existed in the region of the eastern Caribbean. They may represent the core of a subsided lost continent. Dr Heezen observed that the rock formations at the continental slopes on both sides of the Atlantic Ocean do not bend down but break off abruptly as if some terrific force had cracked the continental block and pulled the sides apart.

Dr Heezen also identified coral at extreme depths in the Puerto Rico Trench. Coral reefs do not grow at more than a depth of fifty feet. These reefs dropped thousands of feet into the far reaches of the ocean.

After the analysis of dredge samples taken from the Vema Fault, a long east-west fracture lying between Africa and South America at the latitude of eleven degrees north, fragments of shallow water limestone were found. Minerals in the limestone indicated that they came from a nearby source of granite that was unlikely to occur on the ocean floor. More exhaustive testing of the limestone showed traces of shallow water fossils implying formation in very shallow water. This was also confirmed by the ratios of oxygen and carbon isotopes. One piece of limestone even showed traces of tidal action. It dated back to the Mesozoic period of 70 million years ago to 220 million years ago and forms a cap on a residual continental block left behind. Vertical movements made by the block appeared to have raised it to sea level at some time. Two University of Miami researchers recovered the fragments in 1971. Are these dates correct?

Professor Ewing of Columbia University found evidence that lava had spread only recently, geologically speaking, on the bottom of the Atlantic ocean indicating that the land must have sunk two to three miles or that the sea-level was two to three miles lower than it is now. The sea level has never been as low as this so the land surface must have dropped the two to three miles.

Deep-sea drilling by Ewing and Donn concluded that the Atlantic Ocean warmed up relatively quickly 11,000 years ago at a time when the Northern Ice Sea, which until then had been ice-free, suddenly gained an ice cover.

In 1949 Professor Ewing explored the Mid-Atlantic Ridge. At a depth of 2.5 to three miles he discovered prehistoric beach-sand.

Professor Ewing measured the depth of the thickness of sedimentary ooze in the Mid-Atlantic area and in the foot hills of the Mid Atlantic Ridge where there were thousands of feet of sediment yet on the vast plains that extended either side of it it was less than one hundred feet thick. This area must have sunk only recently geologically speaking.

Granitic rocks which must have been part of a continent were dredged up near the Mid Atlantic Ridge from a depth of 3,600 feet. The rocks had deep scratches on them similar to stones in drift formations normally attributed to the action of glaciers. In the same area were loosely consolidated mud stones that were so soft and weak that a glacier would have destroyed them. These rocks and mud stones are not supposed to be here. The mud stones could only have been formed out in the open air.

The Faroes Islands in the Atlantic Ocean are all that remain above sea level of previously large landmasses.

The Beerenberg Volcano on the island of Jan Mayen in the Atlantic Ocean is all that remains above sea level of a previously much larger landmass.

Many of the undersea valleys in the Atlantic Ocean are actually continuations of existing rivers. This proves that the sea bottom in places must have been above sea level.

Before you start saying that these events may have occurred millions of years ago, the reason that they are in this section is because they are all dated at around eleven thousand years ago.

Examinations of the beds of continental shelves of the Atlantic showed that the beds of rivers that flowed into the Atlantic continued right on out along the shelf, sometimes going through canyons as rivers often do, in France, Spain, North Africa and North America. They continued until they reached an average depth of 1.5 miles.

There are also submerged river canyons on Continental Shelfs from the Rhone River, the Loire and the Seine in France, the Tagus in Portugal as well as from now nonexistent rivers in North Africa. A bulletin of the Geological Society of America commented that these sunken river canyons suggested that worldwide lowering and rising of the sea levels amounting to 8,000 feet must have occurred since the late Tertiary age, namely the Pleiocene Era, the age of man. Sea levels have never done this so the river canyons were once above sealevel until they suddenly dropped.

With sea levels reduced to their Ice Age levels there would be extensive lands reaching out thirty miles or more into the Atlantic from Portugal, Spain and Morocco. This would give in excess of 8,000 square miles of habitable coastal plains in the Gulf of Cadiz and northern Morocco alone. Where should we look for ruined cities then? On the Glacial Period coasts of course.

Some theorists have deduced that the cessation of the Ice Ages in Europe and North America was due to the sinking of the mountain ranges under the Atlantic Ocean. The French Geologist Pierre Termier is of the opinion that a piece of tachylite lava had been able to solidify only in the air but was found at the bottom of the Atlantic Ocean north of the Azores. In 1898 a French cable ship found a piece of vitreous lava or a tachylite at a depth of 1,700 fathoms in the Atlantic Ocean. This tachylite is only formed above sea level. This means that a volcanic eruption occurred here when it was dry land. The piece of tachylite lava came from the Telegraph Plateau near the Rejkijanes Ridge. The plateau derives its name from a dramatic event in the history of the laying of the Transatlantic Cable. The cable laid in 1898 suddenly snapped at 47 degrees North and 29 degrees 40 minutes west of Paris and the ends disappeared into what appeared to be bottomless Sea about five hundred miles north of the Azores. They were only retrieved with great difficulty and during the course of this other items were dragged up from the seabed. The grappling irons drew up soil and broken pieces of rock, which established the fact that the bottom of the Sea at this latitude is mountainous with high summits, steep slopes and deep valleys. This included the huge piece of tachylite lava. The specimen was amorphous, vitreous and of noncrystalline structure indicating that it could only have been solidified in free air and not under water. If it had been formed under 3,000 metres of water it would have become crystallized. It could only have been ejected by an above water volcano and not a submarine one when the whole region sunk 6,500 feet at the same time as the eruption or very shortly afterwards. Tachylite dissolves in seawater in 15,000 years. The cable was the Brest-Cape Cod Cable, which had broken 500 miles North of Punta Delgada. In summer 1898 it snapped at 47 degrees North and 29 degrees forty minutes west. The cable had snapped because the seabed had suddenly ascended 1,200 metres and thus damaged the cable. The formation that the cable had snapped on had been above sea level 11,000 years ago, 9,000 BC. The surface of the Atlantic Ocean 562 miles north of the Azores was covered with lava, which was once above water, and the ruggedness of this showed that the sinking of the land to 3,000 metres below sea level followed suddenly on the emission of the lava. This shows that this area of the Atlantic Ocean is still very geologically unstable today. In 1923 in the same area as the 1898 broken cable repair operations were carried out again at the same site. During the intervening years the sea floor here had risen another 4,000 feet.

Dr Maria Klenova of the Soviet Academy of Science examined rock samples dredged up from 6,600 feet, 2,012 metres, five hundred miles north of the Azores and in her opinion the rock was formed at atmospheric pressure approximately fifteen thousand years ago.

Mr Haug concluded after his research that the convex arch formed by the Antilles and the one found along the western border of the Mediterranean were connected in Tertiary times by tangential chains of land. Therefore a coastline stretched across the Atlantic from Venezuela to Morocco and another between the Lesser Antilles and Portugal, the intervening spaces being covered by Sea. Was this in conjunction with lower sea levels? If so, was this how Solutreans from Europe first arrived in South America?

Professor Scharff concluded that Madiera and the Azores up until the Miocene Era were connected with Portugal and that from Morocco to the Canary Islands and from them to South America stretched a vast land which extended southward as far as St. Helena.

Professor Scharff of Dublin University stated that there was a south Atlantic continent that stretched from Morocco in Africa and the Canary Islands and then southwards to South America. The northern part of this landmass remained until Miocene or Late Tertiary times when the Azores and Madeira became isolated from Europe and the North and South Atlantic joined up. In the early Stone Age man could have reached what are now islands by land. And early man could have colonized this lost land with its herds of elephants.

M. Germaine studed the fauna of the Atlantic Islands and stated that their origin was continental. This was indicated by the freshwater *pulmonata mollusca* called *oleacinidae* which are unique to Central America, the West Indies, the Mediterranean basin and the Atlantic Islands. Germaine believed that this landbridge disappeared near the end of Miocene times.

Professor Hans Petterson on the "Albatross" in sounding the bed of the equatorial Atlantic discovered traces of freshwater plants at a depth of over two miles. Land must have suddenly sunk there.

So there is no evidence for a sunken land mass and geological instability along the Mid-Atlantic Ridge?

John Albert Foex feels that the low water level after the Ice Ages, 450 to 500 feet below present levels, would expand the continents to the limits of the continental shelves. The Bahamas would encompass the Great and Little Bahama Banks and the Azores and Canaries would be enlarged.

Ternier stated that land located nine hundred kilometers north of the Azores Islands in the Atlantic Ocean was plunged into the sea in comparatively recent times geologically speaking.

Nothing unusual happened in this period?

Australia, Niugini and Tasmania were all one continent up until twelve to ten thousand years ago. Species such as the duckbilled platypuss and the spiny anteater cannot tolerate saltwater and are found in Tasmania, the large island to the south of the Australian continent as well as on the mainland. To the north Nuigini and northern Australia were joined together. Young upraised terraces on the Huon Peninsula in Nuigini indicate this as well as submerged channels created by now vanished rivers on the bed of the Arafura Sea. The Gulf of Carpentaria was then Lake Carpentaria. Remember that sea levels before this period were three hundred feet lower than they are today because so much water was locked up in the glaciers and ice masses.

Up until the end of the Quaternary Epoch there was a large landmass in the Atlantic Ocean in the area of the Azores Islands that covered an area of 154,400 square miles (400,000 kilometres) and that suddenly sank about two miles (3 kilometres). This was between 12,000 and 20,000 years ago, 10,000 BC to 18,000 BC.

Professor Hull believes that the Azores Islands are the remains of the peaks of a submerged continent that flourished in the Pleistocene Period. The flora and the fauna of the two hemispheres support the geological theory that there was a common centre in the Atlantic where life began and that the time that this great Atlantean continent existed there was also a great Antillean ridge in the neighborhood of the West Indies. At this epoch the British and continental rivers flowed out many miles past their present outlets and the mid-Atlantic island enjoyed an equable climate when the British Isles was of a semi-Polar climate. During and prior to the Glacial Period great land bridges north and south spanned the Atlantic. Professor Hull also concluded that at this time the Great Antillean ridge shut off the Caribbean Sea and the Gulf of Mexico from the Gulf Stream. The Azores are the tops of very tall mountains, over twenty thousand feet above the seabed plains.

M. Ternier states that this part of the Atlantic Ocean is a great volcanic zone with the eastern part near the Azores being particularly unstable and only submerged quite recently in geological terms. This is based on the mountainous nature of the Azores and the fact that lava and volcanic detritus from the sea area between Cape Cod in Maine and Brest in France north of the Azores must have cooled quite recently under atmospheric pressure or above water.

Deep-sea soundings made in the mid Nineteenth Century indicate that the Azores islands are the mountaintops of a colossal mass of sunken land and that from this centre one great ridge runs southward for some distance and then bifurcating sends out one limb to the shores of Africa and the other to the shores of South America. There are evidences that a third great ridge formerly reached northwards from the Azores to the British Isles. The Azores is a group of volcanoes built upon a foundation of limestone similar to those of Mediterranean countries.

Along the Azores-Gibraltar ridge numerous remains of elephants have been found in at least forty different sites. Some of these sites were at depths of only 360 feet below sealevel. Elephant tusks were found in submerged shorelines, sandbanks and shorelines that were once above sealevel as well as other places that were all originally above sealevel as recently as 25,000 years ago and possibly much more recently. The elephants weren't just going out for an oceanic swim were they? This depth was that of the continental shelves at the end of the last Ice Age. Remember Plato stating that there were elephants in Atlantis?

When sea levels were four hundred feet lower the central islands in the Azores of Pico and Faial, Fayal, would be joined together and most of the others were double in size. Up to ten now unknown islands were in existence that formed an archipelago. This is similar to those depicted on the Piri Reis map that shows seventeen islands instead of the nine shown today. Many of them are fairly large and the present largest island Sao Miguel was the size of Cyprus.

University of Halifax drilling in 1973 and 1974 indicated by rock samples that the rocks at a depth of 800 metres below sea level were above sea level when they were formed.

There are sandy beaches in deepwater plateaus located around the Azores. Sand can only develop at shorelines and not beneath the sea.

When the lost landmass still existed in the North Atlantic the Azores Plateau blocked the way to the European coasts of the warm marine currents. When the landmass sunk the Gulf Stream gradually caused the ice to disappear.

There are freshwater springs still bubbling up from the ocean in the vicinity of the Azores Islands. Azorean fishermen still know where to find fresh water even in the open Sea.

Rocks from the vicinity of the Azores show evidence of great explosions and sudden sinkings in the remote past.

Professor Sharff of Dublin University concluded that up to Tertiary times the Azores and Madeira in the Atlantic were joined to Portugal. Professor Scharff of Dublin University stated that there was a south Atlantic continent that stretched from Morocco in Africa and the Canary Islands and then southwards to South America. The northern part of this landmass remained until Miocene or Late Tertiary times when the Azores and Madeira became isolated from Europe and the North and South Atlantic joined up. In the early Stone Age man could have reached these new islands by land.

Professor Scharff stated that the existence of rabbits in the Azores prior to European contact as well as the presence of 65 species of burrowing beetles called *Amphisbonidae* in America, Africa and the Mediterranean, the fact that identical forms of earthworm are found only in North Africa, the Atlantic islands and Europe, and the marked affinity between forms of a crustacean called the freshwater *decapod* on both sides of the Atlantic indicated the presence of a landbridge across the Atlantic Ocean.

John Albert Foex feels that the low water level at the end of the Ice Ages, 450 to 500 feet below present levels, would expand the continents to the limits of the continental shelves. The Bahamas would encompass the Great and Little Bahama Banks and the Azores and Canaries would be enlarged.

The whole Bahama Banks is a landmass that was dry land as recently as eleven thousand years ago, 9,000 BC. If one were to reduce the sea level by fifty feet then there would be one large island indented by the Tongue of the Ocean up to the side of where Andros Island would be.

The mangled remains of numerous late Pleistocene Era animals have been found on the Balearic Islands in the Mediterranean Sea. The remains were attributed to massive water action or a tidal wave. Was this from the flooding of the Mediterranean Basin? Even with these closer dates still allow for geological dating variance.

There was a dry and ice-free land passage between Siberia and Alaska. Remember that though we have an earlier date for the submergence of Beringia many of these geological estimated ages are purely that, estimates. There is that geological dating variance again. And how many times has this happened as we have seen extreme rises and falls in sea levels in only the last few thousand years.

In the Northern Pacific Ocean a continental landmass that occupied the whole of the Bering Straits and the immediate area south of it suddenly foundered at the end of the Pleistocene Period. This area stretched down to the arc of the Aleutian Islands which extended below what would have been a large inland sea. Was there a large plain between the present Aleutians and what is now the Bering Strait?

An extensive cedar forest grew on the Greater Bermuda Bank, which is now under water. Carbon-14 tests showed that the forest was wiped out about 11,000 years ago around 9,000 BC. This date is close enough to ten thousand BC to be the same period. Or had this forest grown on the Bermuda Bank in the period after the Bermuda Impact around 10,000 BC? Had the Bermuda Crater been formed two thousand years before the second impact event?

The Philippines, Sumatra, Borneo and Java were together connected to the landmass of continental Asia. This now submerged area was called Sundaland.

Unusual remains were found near Santa Lucia in the Lagoa do Sumidouro in Minas Gerais in Brazil. Several caves around the Lagoa do Sumidouro in Brazil contained the bones of more than fifty humans of both sexes and from infants to old people that were mangled and combined with numerous splintered animal bones inside hard clay. These include the remains of *megatheriums*. Some massive force of water had smashed them all together. The skulls of the humans were dolichococephalic or long headed which was distinctive of old European Neanderthals which had been extinct in Europe for 25,000 years. What were they doing in Brazil in South America? Were they probably living there after having travelled across the mid Atlantic continent or island chain? The remains of *megatheriums* and *smilodons*, a giant cat, were alongside the human remains. They apparently all perished together.

Captain Elliot of the U. S. Navy in 1827 discovered numerous human remains embedded in limestone in the area of Santas on the banks of the Rio Santas between the Port of Santas and the town of St Paul. There were hundreds of remains of humans in the calcareous tufa. The presence of *serpulae* or oyster shell in the rock indicated that the burial of the bodies was a marine action though it was far inland. This indicates that a form of tidal wave did this and that it travelled an incredible distance from the Atlantic Ocean. The bodies were lying in the rock in an oblique direction with the heads uppermost and the lower extremities dipping at an angle of twenty to twentyfive degrees below the horizon. Were they swimming? Trying to survive in a massive tsunami? Do we remember that it was postulated that there was a massive lake in the Matto Grosso of Brazil up until this period?

Also in Lagoa Santa in Minas Gerais State in Brazil human skeletons have been found buried under the bones of the *toxodon*, an ungulate creature about nine feet long, the *megatherium*, a huge tardigrade measuring twenty feet and the dinosaur. Those pesky megatheriums again! As well as a dinosaur? The remains of Lagoa Santa are generally dated to 9,000 BC.

We certainly have a mix up of dates here or were these creatures all contemporary and when?

Who is not being extinct when they are supposed to be or are our accepted dates actually out of order due to radiation surges in the past?

In 1970 Canadian Archaeologist Alan Lyle Bryan found a highly mineralized calotte, a skullcap, with very thick walls and exceptionally heavy brow ridges in a paleontological collection from caves in the Lagoa Santa region. The skullcap was of *homo erectus* and was a European skull due to differences in important measurements. It was similar to other skulls found in caves in the Sumidouro cave in the Lagoa Santa region in the 1930s and that more skulls that were the same had been found in the same area in the late 1970s. *Homo erectus* was not supposed to have been in the Americas as only fully formed man was supposed to have come over the Bering Landbridge. Otherwise we have to allow for far greater dates for the existence of man in the Americas. Was there a separate evolution in the Americas or a separate migration?

At high elevations on several islands in Arctic Canada there are numerous sea shells dating from the end of the Paleolithic period. How did they get up there?

Many of the Canadian islands north of Hudson Bay were ice free during the last Ice Age.

Numerous frozen pieces of wood, pine cones and acorns were found on Banks Island in Arctic Canada at an elevation of one hundred and fifty feet above sea level at the centre of the island. One of the trees found there had grown on the spot and still retained its leaves. No oaks grow within several hundred miles of the area now. Eleven conifers were found here as well. There were also found the remains of several former immense forests on the island. Sir Robert McLure found one cliff on the island to be totally composed of one mass of fossil trees in confusion for a depth of forty feet. This was not the only area like this. During the last Ice Age Banks Island was ice free and a form of refuge for flora and fauna that could not exist in much of Canada or the United States due to the glacial ice mass. This could indicate that the Ice Age was not caused by a general dropping of world temperatures but by the new positioning of the poles.

The Old Cordillera or Cascade Culture of the Columbia River Valley in British Columbia in Canada that lasted to 5,000 BC appeared. This culture was a fishing culture. They also used edible plants.

Warm unglaciated climate conditions existed on Melville Island in the far north of Canada.

The Vancouver Peninsula in Canada began rising from sea level after having been buried under one mile of ice.

The Canary Islands in the Atlantic Ocean of Palma, Fuertoventura and Gomera are parts of an older mountain range consisting of diabase, an eruptive volcanic rock. On Grand Canary and Palma an upheaval of from 600 to 1,000 feet can be demonstrated and in Madeira up to 1,400 feet. Professor de Lapparent favoured the existence of a coastline or an island chain during the Miocene Era connecting the West Indies with Southern Europe. The end of the Pleiocene and the whole of the Pleistocene Period were distinguished by a series of subsidences, which resulted in finally opening up the northern depression of the Atlantic Ocean.

Submerged cities have apparently been sited around Boa Vista Island in the Cape Verde Islands in the Atlantic Ocean. In 1969 the remains of a marketplace were actually discovered on the seashelf which was above sealevel prior to the cataclysm of 9-10,000 BC. Cities were not supposed to exist in this period but we seem to have found a few of them by now.

The Bayan Kara Shan Mountain Range was raised two thousand meters or 6,500 feet. At the same time Minya Konka, one of the worlds highest peaks, was lifted one thousand metres or 3,250 feet. At the same time mountains in the Yunnan Ranges also rose two thousand metres. To the south the Tibetan Plateau rose three thousand metres or 9,750 feet. The Bayan Kara Shan Mountains are in Garze in Sichuan in China.

In Chou-kou-tien, also called Zhoukoudian, near Beijing in China in 1933 during excavations Dr. F. Weidenreich discovered a number of skulls and skeletons here. The fractured skulls of seven people were found in a fissure. One skull belonged to an old European, another to a woman with a narrow head, typically Melanesian in character, and a distinctively Eskimo young woman. These skulls are 30,000 years old, 28,000 BC. The three skulls were underneath a great layer of thousands of animal bones dating from the cataclysm of 9-10,000 BC. The caves are crammed with assorted human and animal bones in astonishing diversity, apparently swept there from far away by a titanic flood, or judging by the age differences, by several different floods. Or massive tidal waves during the major cataclysm. The animal remains were of numerous different types of animals from different climatic zones. There were mammoths, buffaloes, ostriches as well as arctic animals amongst others who were all crammed together into the caves and fissures. There was no glacial ice cover in Northern China during the last Ice Age so only rapidly moving tidal waves could have moved them all here. A human skull also found here is 500,000 years old, 500,000 BC. These caves were also where Peking Man was also discovered whose remains are seven hundred and fifty thousand years old.

The Gobi Desert is called the "Sham Ho" in Chinese, the "Dried Up Sea". The Chinese also called the Gobi Desert the "Great Han Hai", the Great Inland Sea. The Great Inland Sea extended from the Great Kingan Shan to the east and the Pamirs and the Tien Shan Mountains to the west. This was a distance of two thousand miles across and from north to south seven hundred miles. Incidentally the entire basin was two to three thousand feet lower than it is today and was raised at the same time as the Tien Shan Mountains, the Pamirs and the Tibetan Plateau. Was the draining away of the waters of this immense sea a contributor to deluge legends in this area? The Gobi Desert was originally a sea.

On the Great Kingan Shan at the edge of the Gobi Sea huge lava outpourings occurred. At the same time there were large basalt flows on the neighboring Sikhote-Alin Mountains and a spectacular collapse of the sea floor by many thousands of metres down off the eastern shoreline of the Kamchatka Peninsula all the way to the south to the Japanese Islands.

There have been huge uplifts of the earth's crust around the Gobi Desert since the end of the Glacial Period according to Dr Bailey Willis in the early Twentieth Century. To the west of the Gobi the Tien Shan Mountains and the Russian Altai Mountains according to him were raised up after the glacial period ended. The present Gobi Basin was formed with the uplift of the TransBaikal Ranges.

How did the Chinese in ancient times know that the Gobi Desert was once a sea?

In the Gobi Desert there are huge masses of rock that have been fused together under intense heat and the rocks have been vitrified by huge heat sources coming from above. A stratigraphic layer of fused green glass has also been found here that could only have been caused by a nuclear explosion. This stratified green glass has been found worldwide in many areas that are incidentally all deserts now! Are these our invisible heat sources that we have already met that occurred in conjunction with the cosmic bombardments or nuclear explosions? Both could have caused the vitrified green glass.

The Tien Shan Mountains in China rose as the Gobi Sea was suddenly raised up two to three thousand feet.

The Yunnan Ranges in China suddenly rose two thousand metres. There seem to be a lot of reports of mountain ranges around the world suddenly rising in this period! Were a lot of different geologists colluding together?

Cataclysmic geology is now making a comeback as well. Gradualism did not have all of the answers after all.

There was massive and sudden extinction of animals in Denmark in the period of 9,000 to 10,000 BC.

A stratigraphic layer of fused green glass has been found in Egypt that could only have been caused by a nuclear explosion. In the Libyan Desert which is in fact in Egypt hundreds of thousands of strange glass objects have been found. These were pieces of glass from that of the size of a pea to that of an egg and they were scattered over the ground and not buried in it like meteoric fragments or tektites. This was in a sand free area. These objects were made of the same silicone as the desert sand nearby and appeared to have been created by intense heat, in the range of that level of heat produced by a nuclear explosion. This is the same as the fused green glass found in the Gobi Desert and from the same period! Were there invisible heat sources as well as meteorites, asteroids or comets hitting the Earth in this period as well? Had these aerial heat sources helped contribute to the desertification of the area?

The massive Zodiac of Denderah which was found in Egypt begins oddly enough at the Vernal Equinox of Leo and covers the period of 10,950 BC to 8,800 BC. This is supposed to be far earlier than Egyptian civilization. Why would anyone want to do a Zodiac of the past unless the Zodiac is older than first thought? It is funny how familiar the dates in the zodiac seem now. Is it a warning of times past? The Zodiac of Denderah was carved into the ceiling of the temple of Denderah and shows the configuration of stars around 9,000 BC. Was the Zodiac of Denderah actually carved as a memorial to the past as well as a warning to the future? That the events of this period could be repeated?

Joseph Prestwich, formerly Professor of Geology at Oxford University concluded that England, Central Europe, the Mediterranean islands of Corsica, Sardinia and Sicily were all completely submerged on several occasions during the rapid melting of the ice sheets at the end of the last Ice Age. The animals naturally retreated as the waters advanced deeper into the hills until they found themselves embayed. They thronged together in vast multitudes, crushing into the more accessible caves until overtaken by the waters and destroyed. Rocky debris and large rocks from the sides of hills were hurled down by the currents of water, crushing and smashing the bones.

Wide plains extended from the present area covered by the English Channel into the Atlantic Ocean.

At Brixham near Plymouth in South Devon in England there are three bone caves wherein the smashed and mangled bones of rhinoceros, bear and deer have been found jammed together as if by a massive force.

More mangled animal bones were found in Ash Hole Cave in Berry Head in Devon in England. The smashed and mangled bones of numerous unrelated animals have been found jammed into the cave system here by some unknown but huge force. The remains of animals were found piled in confusion against the sides of the cave and jammed into the interstices. More tsunamis?

On top of the Norwich Crag in England are the remains of an ancient forest showing stumps of trees standing erect with their roots penetrating into ancient soil. In this soil are the remains of many extinct animals together with others that are still living. These include the hippopotamus, three species of elephant, the mammoths, rhinoceros, bear, horse and Irish elk. How and why had they all congregated here all of a sudden? Or were they pushed here?

There are submerged forest deposits in the sea off the Pentewan Valley in Cornwall in England that were found during tin mining work.

More mangled animals were also found in Kent's Cavern in Torquay in Devon in England. The smashed and mangled bones of numerous unrelated animals have been found jammed into the cave system here by some unknown but huge force. The remains of animals were found piled in confusion against the sides of the cave and jammed into the interstices.

At Oreston near Plymouth in South Devon in England there are three bone caves wherein the smashed and mangled bones of rhinoceros, bear and deer have been found jammed together as if by a massive force.

The remains of hippopotamus, mammoth, rhinoceros, horse, bear, bison, wolf and lion have been found in the neighborhood of Plymouth on the English Channel dating from the end of the last Ice Age. The bones are all shattered and incomplete and incredibly jammed together.

More smashed up animals were found on the Isle of Portland in Dorset in England. The smashed and mangled bones of numerous unrelated animals have been found jammed into the cave system here by some unknown but huge force. The remains of animals were found piled in confusion against the sides of the cave and jammed into the interstices. These cave jammings were not caused by stampedes but by being forced into position by some incredible wave action.

Man was living in the valley of the lower Thames River in England before the Arctic mammalia had taken full possession of it and before the big-nosed rhinoceros had become extinct.

The Thames and the Humber have been found to empty into the former Rhine River under the North Sea which at the time was countryside.

At the bottom of the English Channel the remains of a long submerged river valley have been found. River Channels cannot form under water. Was a river all that was separating England from France?

There are submerged river canyons on Continental Shelfs from the Rhone River, the Loire and the Seine in France, the Tagus in Portugal as well as from now nonexistent rivers in North Africa.

The Rhine River valley continues up the North Sea midway between South Norway and Scotland indicating that this was dry land in recent times. The Rhine River runs across what is now the North Sea and empties into the

Atlantic Ocean near the Orkney Islands. Some other authorities state that it does this near Aberdeen in Scotland. Stone Age tools and mammoth teeth have been found dredged up from the shallow bottom of the North Sea. This indicates that it was once coastal land.

A bulletin of the Geological Society of America commented that these sunken river canyons suggested that world wide lowering and rising of the sea levels amounting to 8,000 feet must have occurred since the late Tertiary age, namely the Pleiocene Era, the age of man. Plains generally extended all along the coastlines of Spain, France, Italy, and Greece where many of the islands were originally joined together.

A large landmass broke up and became submerged in this period. Fennoscandia was a northern landmass that connected Spitzbergen to northern Europe and Asia. Fennoscandia sank into what is now the Arctic Ocean and included much of Northern Europe and the numerous archipelagos now in the Arctic Ocean up to what is now only five hundred miles from the North Pole. The remains of immense forests have been found here of alder, elm and oak that reached far into what is now the Arctic and down to the Ural Mountains and Siberia. Today nothing grows here, especially trees and bushes.

There are submerged river canyons on continental shelfs from the Rhone River, the Loire and the Seine in France, the Tagus in Portugal as well as from now nonexistent rivers in North Africa. River valleys do not form under water.

Plains generally extended all along the coastlines of Spain, France, Italy, and Greece where many of the islands were originally joined together.

Fissures in the rock on top of isolated hills in central France are filled with what is known as osseous breccia consisting of the splintered bones of mammoths, woolly rhinoceroses and other animals. What had splintered the bones en masse in this period?

The drift gravels of France were deposited by violent cataclysms. The gravel beds in which the remains of man and extinct animals are found lie at an elevation from eighty to two hundred feet above the present water levels of the valleys. Had gigantic tsunamis caused these?

A large part of Aquitaine was submerged during the Cataclysm.

A large part of Brittany was submerged during the Cataclysm.

Eleven miles northwest of Quimper is the supposed site of the lost city of Ys in the bay of Doaurnenez. There have been sightings of submerged ruins near Le Ris. Cyclopean ruins of port facilities have been seen in the sea near here on the seashelf that was above sealevel before the period of 9-10,000 BC. Douarnenez is in Finistere in Brittany in France.

There was rapid extinction of massive numbers of animals in Chalon-sur-Saone in Burgundy in France.

Between Dijon and Lyons in Chalon-sur-Saone there is a flat topped hill called Mont de Sautenay. In a rock on the summit there is a fissure filled with an incredible variety of shattered animal bones from oxen, wolves, bears, horses and others who all appear to have ascended this singular and lone hill to escape from something but instead were crushed into the fissure. None of the bones appeared to have been gnawed or chewed and they all apparently died together. Was this another tidal wave?

The 1,430 foot peak of Mount Genay is capped by a breccia containing the remains of mammoths, woolly rhinoceroses, horses, reindeer and other animals. Mount Genay is in Cote-d'Or in Burgundy in France. Had the unfortunate animals congregated here so as to be safe from massive flooding? Had they been proved wrong? Had they been pushed up there by tsunamis?

The strata at St. Acheul are composed of an upper strata representing brick earth four to five feet in thickness and containing a few angular flints. The next is a thin layer of angular gravel, one to two feet in thickness. The next is a bed of sandy marl, five to six feet thick. The lowest deposits overlay the chalk and are a bed of partially rounded gravel in which human implements have been found along with the bones of mammoth, the wild bull, the deer, the horse, the rhinoceros, and the reindeer. All land dwelling animals with no marine deposits. It is at the lowest part of these deposits that the flint implements occur. St Acheul near Amiens is in the Somme of France.

More relics were found in St Prest near Chartres in Eure-et-Loire in France. In the upper Pliocene strata stone implements have been found that are for cutting on bone in connection with relics of a long extinct elephant, *Elephas meridionalis* that is wholly lacking in later strata of drift.

The Rhone River Delta in France is the site believed to be the source of a chevron or megatsunami eleven thousand years ago. The only possible object that could have created this megatsunami is an impact by an asteroid or meteorite as there are no other explanations from the geology of the area which is now submerged. The Mediterranean Sea has been filled and emptied several times.

Once again more evidence of smashed animals and plants being suddenly jammed into caves and fissures around the same time.

A stratigraphic layer of fused green glass has been found in Pierrelatte in Gabon in Africa that could only have been caused by a nuclear explosion. Also at Pierrelatte there is the Oklo Mine where the uranium ore of the mine is so depleted that it only contains abnormally low proportions of U235 that are normally only found in depleted uranium fuel from nuclear reactors. The mine also contained four rare elements in forms similar to those found only in depleted uranium. This seemed to be a natural and ancient nuclear reaction or was it?

The Rhine River valley continues up the North Sea midway between South Norway and Scotland indicating that this was dry land in recent times.

The Rhine River runs across what is now the North Sea and empties into the Atlantic Ocean near the Orkney Islands. Some other authorities state that it does this near Aberdeen in Scotland. Stone Age tools and mammoth teeth have been found dredged up from the shallow bottom of the North Sea. This indicates that it was once coastal land.

In Europe the North Sea did not exist, most of the land being covered with a mile thick Ice Cap.

There was also a massive and sudden extinction of animals in Germany in this period.

Yes, I know that a lot of data is repeated but this is to satisfy the tablet jumpers who want to go straight to where they want to go and get all of the data in connection with it. This then is the compromise and in your case a revision.

The Straits of Gibraltar between Spain in Europe and Morocco in Africa was originally a narrow channel sixty miles long with two small islands at its Atlantic entrance.

Some three hundred miles due west of the Straits of Gibraltar along the present day site of the Gorringe Ridge was a larger island the size of present day Menorca. This is now beneath the sea but parts of it even today are only as little as 65 feet below the surface. There are two sunken seamounts just beyond the Straits of Gibraltar on the edge of the Continental shelf that would have risen out of the sea when sea levels were considerably lower. Are these the real Pillars of Hercules or Pillars of Melkart as the Phoenicians called them?

The rapid rise in sea levels at the end of the last Ice Age would have devastated the wide plains bordering the narrow Gibraltar Channel.

The Mediterranean was a much smaller inland sea during the Holocene Era and supported a population on what is now sea bottom. The Atlantic Ocean broke through the Straits of Gibraltar, the Pillars of Hercules, flooding and filling the Mediterranean Basin around 10,000 BC to 9,000 BC. Yes, these are repeated but so many geologists, so many dates. You thought it was a science did you?

On the Rock of Gibraltar amongst the scrambled animal bones of the last Ice Age a human molar and some flints were discovered. Broken pieces of Neolithic pottery were also found within these remains. Man was not supposed to be creating pottery in this period unless he was in Japan and some other places as well. The Rock of Gibraltar is extensively fissured and filled with the mixed and mangled bones of many different types of late Pleistocene animals including panther, lynx, wolf, bear and many other types of animals. The bones have been broken into thousands of fragments and this could have only been caused by a massive and sudden flood that drove the animals into the fissures and the crags. Marine shells and coral were also found in the fissures some down to 290 feet. The mud that encased them has become a hard breccia. The Rock of Gibraltar is riddled with hundreds of caves. The limestone caves are famous for their limestone formations.

Plains generally extended all along the coastlines of Spain, France, Italy, and Greece where many of the islands were joined together. The Mediterranean was much smaller at the end of the last Ice Age.

At the end of the Quaternary Epoch, eleven thousand years ago, huge masses of water had been converted into sheets of inland ice and the sea level was 330 to 660 feet lower than it is today. This is the eustatic fall in sea level. The Quaternary coastline of Attica in Greece was really like a promontory jutting far out to sea from the mainland. In the region of Athens it was several times as wide and was joined to Euboea and the Peloponnese by land bridges. The mountains appeared 330 feet higher than today so that the Acropolis was 800 feet high instead of 500 feet high today. Plato described the flat expanses that now form part of the sea as fertile plains. Plato was a liar? He also described elephants on Atlantis. We have already found them.

Greece during the high Glacial Period was markedly different from the Greece of today. There were huge coastal plains that today are very rare. These lowlands were inundated when world sea levels rose after the collapse of the glaciation. Evidence of this was found in Larisa in Thessaly.

On the island of Cerigo, now called Cythera, in the Ionian Sea the fissures on the sides of a barren truncated mountain one mile in circumference are filled with the remains of a multitude of animals from the end of the Pleistocene Era. The mountain is locally called the "Mountain of Bones" and is covered from base to summit in these bones. There is no explanation for how all of these different bones were jammed into this mountain unless we allow for the sudden onrush of a tidal wave of once live animals now all jammed together.

There were large trees in Greenland such as fir, spruce, marsh cypress, hazel and numerous others. Now no trees can grow there. This indicates massive climate change in this period.

The Wyville Thomson Ridge between Iceland and the Orkney and Shetland islands was originally a large shelf area that has now dropped from 1,000 metres to 2,500 metres below sea level only recently geologically speaking. Large quantities of shells and other otoliths have been found on the shelf indicating that the shelf must have dropped suddenly or these shallow water sea creatures would have had time to escape to higher ground so to say instead of dying in vast numbers. The area must have dropped 2,000 metres very suddenly. Between Iceland and Jan Mayen there are masses of these shallow water otoliths.

Iceland was four times larger than at present prior to the last Cataclysm. Fjords, narrow trenches or drowned valleys are as much a feature of Iceland as they are of Norway.

The climate in Iceland was mild in this period. In clays from the late Pleistocene period there are found buried large trees such as oak, maple, sequoia and spruce amongst others. The mean average temperature was eleven degrees before they were suddenly interred inside the clay. This part of Iceland was not glaciated in this period. These trees do not grow here now. Now the most common tree native to Iceland is the northern birch which formerly formed forest over much of Iceland along with aspen, rowan and common juniper and other smaller trees.

The Himalayas suddenly rose over three thousand metres or 9,750 feet around eleven thousand years ago.

There was massive tilting and elevation of lake and river deposits in the Kashmir Valley in India. Pleistocene fossil deposits containing plants and vertebrates were raised to a height of 5,000 feet to six thousand feet. There was dissection of river valleys containing post Tertiary period mammals to depths of three thousand feet as over-thrusting occurred when older Himalayan rocks were thrust upon later Pleistocene strata. In Kashmir 5,000 feet of marine beds containing Paleolithic remains was dragged up. The Pir Panjal Mountains in Kashmir rose over 6,000 feet. This upheaval folded and tilted the Karewas and also affected the rocks of the Potwar Range and the Salt Range.

At the end of the Pleistocene period the Gangetic Trough in India was formed when the Himalayas and the Tibetan Plateau were raised to their present height. This trough is 1,200 miles long and averaging 250 miles wide and over 6,200 feet deep. The Ganges River now flows along it. Pleistocene debris here is 6,500 feet deep.

Professor Cesare Emiliani, a marine Geologist at the University of Miami in Florida referring to work done on core samples taken by the research vessel "Glomar Challenger" discovered that a fifteen hundred mile long ridge under the Indian Ocean, now over one mile deep, had once been above water. There had been an island chain here that contained swamps and lagoons. Other deep sea cores with recent deep Sea sediments in their upper levels also contained sediments usually found in shallow water and even dry land.

In 1947 the Swedish survey ship "Albatross" sailed over a vast and continous submerged plateau of hardened lava that filled the earlier valleys on the seafloor making it almost level several hundred miles southeast of Sri Lanka. This area of land would have to be above sea level in late Paleolithic times or the end of the Ice Ages.

Wallace stated that there was a massive submergence of land masses in the Indian Ocean that he called the Great Southern Continent and that this submergence occurred at the end of the global catastrophe around 9,000 BC when it was accompanied by massive volcanic occurrences.

H. F. Blandford states that the Seychelles, Mauritius, the Adas Bank, the Laccadives, Maldives, the Chagos Islands and the Sayha de Malha are all remants of a Great Southern Continent in what is now the Indian Ocean. Interestingly enough Sri Lanka was separated from India around this same time and could have also been part of this lost continent.

A stratigraphic layer of fused green glass has been found here that could only have been caused by a nuclear explosion. This was along the Euphrates

River Valley. We have met a few of these strata of melted and fused glass from this and the previous millennium already.

The Natufian culture in Israel was using sickles in 9,000 BC to harvest grain in what is now desert.

Mastodon teeth have been found from depths of three hundred feet from the bottom of the Inland Sea in Japan. Were they all out swimming? Or was this another now submerged land or another tsunami?

Libya, the ancient name for North Africa was devastated by severe earthquakes during the cataclysm of 9,000 BC.

The great inland sea in Libya, Lake Trithonis, suddenly dried up at the same time.

Vitrified glass has been found in Libya as well from this same period. This is the same type of glass as in Egypt, Iran and Iraq. What sort of aerial heat sources were around in this period? What sort of Hell was this?

In early March 1979, USSR Scientists from the Soviet Academy's Institute of Oceanography located ruins on the seabed 320 to 480 kilometres offshore between Portugal and Madeira. Eight photographs were taken by cable camera from the ship Vitiaz of vestiges of great staircases and walls.

Professor Sharff of Dublin University concluded that up to Tertiary times the Azores and Madeira in the Atlantic were joined to Portugal. Professor Scharff of Dublin University stated that there was a south Atlantic continent that stretched from Morocco in Africa and the Canary Islands and then southwards to South America. The northern part of this landmass remained until Miocene or Late Tertiary times when the Azores and Madeira became isolated from Europe and the North and South Atlantic joined up. In the early Stone Age man could have reached these islands by land.

M. Ternier states that this part of the Atlantic Ocean is a great volcanic zone with the eastern part near the Azores being particularly unstable and only submerged quite recently in geological terms. This is based on the mountainous nature of the Azores and the fact that lava and volcanic detritus from the sea area between Cape Cod in Maine and Brest in France north of the Azores must have cooled quite recently under atmospheric pressure or above water.

There was a massive lowering of sea levels by at least fifty feet up to three hundred feet in the areas of Mount Ophir and Kedah Peak in Malaysia. Was this due to an uplift of the land or a drop in sea level?

Malta was connected to Sicily which was connected to Italy around 9,000 BC. The island of Malta was once connected to the mainland. Originally in the dry valley of the Mediterranean were several huge lakes. The western Mediterranean was separated from the eastern part by a submerged ridge on which the Maltese islands stand. This now hidden land once joined Europe to North Africa and on either side of it were two Great Lakes. The land bridge which connected Africa with Spain at the Straits of Gibraltar was broken through and the ocean roared in, first flooding the western lake, then overrunning the land between Sicily and North Africa, marooning small islands like Levanzo, Malta and Gozo, and then the eastern lake.

During the last advance of the Pleistocene age the Mediterranean was a great low plain with a pair of large lakes separated by the ridge connecting Italy, Sicily and Malta with Africa. During the warm interglacial periods apelike men hunted pigmy elephants and pigmy hippopotami in Sicily. With the final melting of the ice caps the ocean rose until about 15,000 years ago, 13,000 BC, it broke through the Isthmus of Gibraltar and filled the Mediterranean Basin to its present level forcing thousands of men and animals to flee.

Yes, I know that this has been repeated again. Get the geologists to agree and it would not happen.

The skeletons of miniature elephants found in the Ghar Dalam, Cave of Darkness, in Malta suggests that Europe and North Africa were once joined. Three species of mammoths have been found and are distinguished according to size, the smallest being only three feet high. Similar though not identical dwarf elephants have been found in Sicily, Sardinia, Crete and Cyprus. The smaller ones are later than the larger ones, as they are found higher up in the stratified deposits. The remains of pigmy hippopotamuses were also found along with the bones of deer, bear, fox and other animals no longer found on Malta. These are all continental animals.

The Mediterranean was a much smaller inland sea during the Holocene Era and supported a population on what is now sea bottom. The Atlantic Ocean broke through the Straits of Gibraltar, the Pillars of Hercules, flooding and filling the Mediterranean basin in 9,000 BC to 10,000 BC. Other sources say thirteen thousand years ago.

At the end of the Pleistocene era Corsica, Sicily and Sardinia were joined together.

Dwarf elephants have been found in Sicily, Sardinia, Crete and Cyprus. The smaller ones are later than the larger ones, as they are found higher up in the stratified deposits. The remains of pigmy hippopotamuses were also found along with the bones of deer, bear, fox and other animals no longer found on Crete. The island of Crete is honeycombed with limestone caves that like many Mediterranean islands are jammed with the mangled remains of a myriad of late Pleistocene Era animals. The extinction of these animals all over the island is attributed to sudden widespread water action, namely tidal waves.

There are huge crustal displacements from the late Pleistocene Period in central Westland in New Zealand. Vertical crustal uplift east of the Great Fault in the Southern Alps in South Island is estimated to have been as much as 58,500 feet or 18,000 metres with a similar horizontal displacement. A crustal displacement this size would have caused worldwide disturbances. This displacement was upward from the sea bottom.

In North America there was another ice advance into the Superior Basin which blocked off the Saint Lawrence Basin again and a new lake was formed.

In Europe the North Sea did not exist, much of the land being covered with a mile thick Ice Cap. Where there was no ice mass the North Sea was a vast plain that had major river valleys running across it that suddenly submerged. Stone Age tools and mammoth teeth have been found dredged up from the shallow bottom of the North Sea. This indicates that it was once coastal land.

Spitzbergen was inhabited by primitive man. Fragments of prehistoric cliff drawings have been found that show incised outlines of whales and deer near Ny Alesund. Deer are plains animals. The area now is uninhabitable.

Dr von Ihering stated that the Polynesian Islands in the Pacific Ocean were actually the remains of a continental landmass and not sporadic volcanic eruptions. This theory tallies with Poleynesian legends of the lost continent of Havaiiki or Hawaiki which was their ancestral fatherland. The Maori Tare Watere Te Kahu of the Ngai-Tahu people told Mr. S. P. Smith about Hawaiki-Nui which was a mainland, Tua Whenua, with vast plains on the side towards the sea and a high range of snowy mountains on the inland side and through this country ran the river Tohinga.

The Karakoram Range on the border of Pakistan and China rose thousands of feet due to a sudden crustal displacement around 9,000 BC.

At the end of the Meiocene Era the Isthmus of Panama appeared above water which let animals cross over it from North to South America and vice-versa.

On the Peruvian coast near Lima evidence has been found for the cultivation of beans and peppers dating back to the ninth millennium BC. The oldest fibrework in the Americas has been found here as well. Containers made by twisting, looping and knotting plant fibres were found dating to this period. Textiles, wood and leatherwork have also been found. When did man first come to South America? The textiles were found in Guitarrero Cave near Callejon de Huaylas in Yungay province in Peru.

Doctor James A. Westlake, an English Geophysicist, discovered several tektites in the Andes of South America. They resemble fused glass that has been welded together. Two of the three that Dr Westlake found contain rare metallic substances that are unknown on Earth. Dr Westlake, along with several Russian Scientists, suggested that an unknown planet exploded and showered earth with fragments of itself. Tektites from this period have been found from all over the world at this time.

The Philippines, Sumatra, Borneo and Java were together connected to the landmass of continental Asia. This now submerged area was called Sundaland. The dates for the submergence of Sundaland are also in dispute. The only thing not in dispute is the former existence of Sundaland.

The North African savannah with Lake Trithonis at its centre, which was once lush pasture land began drying up at an incredibly fast rate to become the

Sahara Desert. Remains of animals and rockdrawings prove that this was a very fertile area until something unforeseen happened.

Until 2,550 BC the Sahara Desert contained many freshwater lakes including some quite large ones. Crocodiles and hippopotami existed in northern Mali. Lake Chad and other lake basins supported rich plant communities and were full of fish.

In Scotland tons of fossilized fish have been found in positions of terror with their fins extended and eyes bulging. Are these more stampeding fish similar to those in California?

A stratigraphic layer of fused green glass has been found in Scotland that could only have been caused by a nuclear explosion. These layers of fused green glass from this period have been found in several areas. Were they caused by atomic explosions or heat explosions from the sky?

There was sudden and massive extermination of animals in Scotland as well.

About four miles from Loch Lomond an antler was found associated with marine shells near the bottom of a bed of blue clay and close to the underlying rock. The blue clay was covered with twelve feet of rough stony clay. This was in the Endrick Valley in Argyll and Bute.

Remains of a mammoth have been found where they occurred in a bed of laminated sand underlying the till near Airdrie at Chapel Hall in North Lanarckshire in Scotland.

There was a spectacular collapse of the sea floor by thousands of metres along the eastern edge of the Kamchatka Peninsula in Siberia that reached as far south as the Japanese Islands. What caused this massive collapse? Was this a result of the Russian asteroid impacts in this same period or unknown Siberian ones?

Traces of a Stone Age settlement have been found on the Novosibirsk Islands including bone implements as well as arrowheads, needles and axes fashioned from mammoth tusks. This indicates much warmer temperatures in this period than now.

The Pamir Mountain range was raised from being a plain and the remains of reindeer, plains dwelling animals, were found at heights that reindeer could never have lived at.

There were large and unusual basalt flows on the Sikhote-Alin Mountains in Siberia.

Extensive lands extended almost one hundred miles south from the tip of South Africa before they were inundated.

The Atlas (Morocco) and Spanish peninsulas collapsed under a massive onrush of sea tides of colossal proportions, thus creating the Mediterranean Sea. The two peninsulas had been joined together leaving a large landlocked lake in the middle of the Mediterranean that was flooded when the peninsulas collapsed.

Sri Lanka was separated from India around 9,000 BC. Satellite photos show a narrow isthmus linking Sri Lanka to India that is just under the sea. Legends of the area refer to Adam's Bridge linking the island to the mainland. Space images taken by NASA reveal an ancient bridge or causeway in the Palk Strait between India and Sri Lanka. The bridge is made of a chain of shoals approximately eighteen miles or thirty kilometers long. The bridge's unique curvature and composition by age indicates that it is man made. Archaeology states that man first made his appearance in Sri Lanka approximately 1,750,000 years ago and oddly enough the bridge is the same age. The legend the Ramayama was supposed to have taken place in the age of Treta Yuga that is described as occurring in this same period. The Ramayama mentions a bridge that was built between Rameshwaram (India) and the Sri Lankan coast under the supervision of a dynamic and invincible figure called Rama. The land bridge called Adams Bridge is a narrow ridge of sand and rocks that connects Mannar Island in Sri Lanka with Pamban Island in India. At high tide it is covered with four foot of water. Argue about the dates but you still have a very long curved bridge between India and Sri Lanka. There is still controversy over when the the isthmus was submerged with local tradition stating that this occurred in the fifteenth century AD.

People who lived in the Spirit Cave near Pang Mapha which is near Mae Hong Son in Thailand in 9,000 BC were growing domesticated beans, peas, gourds and water chestnuts.

The Himalayas suddenly rose by three thousand metres or 9,750 feet. This is accepted as being at the end of the Pleistocene Period. The Kailas Range in Tibet suddenly rose to a greater height of thousands of feet due to a massive crustal displacement. This was post glacial.

Huge temperate watered plains stretched up to 120 miles out into the Mediterranean from Tunisia in North Africa.

The North African savannah with Lake Trithonis at its centre, which was lush pastureland began drying up at an incredibly fast rate to become the Sahara Desert. Remains of animals and rock drawings prove that this was a very fertile area until something unforeseen suddenly happened.

This is a dried up marsh section of Tunisia, once a bay of the Mediterranean and later the inland Trithonis Sea, Lake Tritonia, with a citadel island in the middle. The central plain of Tunisia was once an island.

Several classical writers mention that Lake Trithonis lost its waters when the dykes burst during an earthquake and eventually dried up becoming the Shott el Djerid, a marshy shallow lake.

In Algeria and Tunisia geodetic surveys made by the French Government showed that the Shotts, shallow marshy lakes were below sea level and would fill up with water if a series of protective sand dunes were removed between them and the Mediteranean Sea. Interestingly enough diggings here found the remains of sea mollusks. How did they get here?

At Shott el Djerid in a swamp the remains of a large town with concentric circular walls was found enclosing a central palace. The area of Shott el Djerid was also locally called Bahr Atala, the Sea of Atlas. Atala was a Central American name for Atlantis. Just a coincidence?

A stratigraphic layer of fused green glass has been found in south-central Turkey that could only have been caused by a nuclear explosion. Here are those strange vitrified spots again from this same period. Were they caused by nuclear explosions or cosmic bombardments?

Most of the present undersea continental shelf off the coast of the United States was dry land about 9,000 BC. Fishermen dragging the sea bottom for scallops and clams have found the teeth of extinct mammoths and mastodons up to 190 miles out to sea and to depths of 400 feet. This was the precataclysmic coast line. The remains of horses, tapirs, musk ox and giant moose have also been found on the continental shelves.

There is also evidence of sunken shorelines, sands and deposits of peat, which has led scientists to conclude that in 13,000 BC the United States continental shelf was a wide coastal plain teeming with wildlife and covered with forests. After 9,000 BC it was the seafloor.

The bulk of the animal extinctions took place between eleven thousand BC and nine thousand BC when there were violent and unexplained fluctuations of climate. Geologist John Imbrie stated that there was a climatic revolution around eleven thousand years ago, 9,000 BC. There were also increased rates of sedimentation and an abrupt temperature increase of six to ten degrees centigrade in the surface waters of the Atlantic Ocean. This increase in water temperatures alone would have brought about the end of the Ice Ages.

The Hudson Canyon is an extension of the Hudson River and extends through underwater cliffs almost 200 miles to the edge of the Continental shelf. The channel created by the Hudson River goes out ninety miles but has been found to be three hundred miles long. When was this now flooded shore line drowned?

The Baltimore Canyon was probably cut by the Delaware River in ancient times, as was the Norfolk Canyon by the Susquehanna River.

Dr Bruce Heezen states that the eastern coastline of the United States 11,000 years ago was some one hundred miles farther out in the Atlantic Ocean than it is today. Suddenly in the same period the Ice Age was over and billions of gallons of snow and ice poured into the sea. The result was a dramatic, sudden and terrifying rising of the sea level all around the world. Every continent in this era felt the impact of the ocean's rise as the great melting of the glaciers occurred. The sea levels before this were perhaps three hundred feet lower. Three hundred feet again! It is amazing how many of these independent researchers are validating each other even if they didn't know it themselves.

Deep-sea drilling by Ewing and Donn concluded that the Atlantic Ocean warmed up relatively quickly 11,000 years ago at a time when the Northern Ice Sea, which until then had been ice-free, suddenly gained an ice cover.

Around this date the ice blocking the St Lawrence River suddenly melted. The level of the Great Basin Lakes suddenly fell and the rapid ice retreat opened the northern drainage systems of the Great Lakes.

In North America there was another ice advance into the Superior Basin which blocked off the Saint Lawrence Basin again and a new lake was formed.

These are all independent reports in each of these paragraphs. How do we explain these two conflicting reports? Were they at different times in the one millennium?

The shells of freshwater oysters normally found in tidal estuaries or lagoons have been discovered at different sites off the Atlantic coast of the United States. They are dated at 9,000 BC. The seas had stabilized by 7,000 BC.

Ancient twigs, seeds, pollens, and peat deposits have been found off the Atlantic coast of the United States. These were submerged around 9,000 BC.

The well-known American Archaeologist Frank C. Hibben who coined the term Clovis Point for a certain type of spearhead estimated that 40 million animals were exterminated in North America alone at the close of the last Ice Age.

There were elephants roaming the plains of North America and South America prior to twelve thousand years ago, 10,000 BC, when they suddenly became extinct.

All over North and South America Ice Age fossils have been unearthed in which incongruous animal types, carnivores and herbivores, are mixed promiscuously with human bones. There are also widespread areas of fossilized land and sea creatures mingled in no order and yet entombed in the same geological strata. It was as if massive tidal waves had overtaken them! We now realize that they had.

In Alabama in the United States Ice Age marine features are present along the gulf coast east of the Mississippi River, in some places at altitudes of two hundred feet. Are these the result of more tsunamis?

The Curtis Lake crater near Palmer in Alaska is 1.6 kilometres in diameter and eleven thousand years old. Is this crater related to the craterlike shallow depressions from the period of 9,500 BC in the area of Point Barrow in Alaska? These are the same type of craters found on the Atlantic coastal plain and the Carolina Bays. There are similar shallow craters in Bolivia and Holland from the same period. These shallow craters in some cases are regarded as being caused by ice masses hitting the Earth such as found in comets. These are oriented shallow depressions or lakes in the permafrost near Point Barrow that are aligned northwest to southwest. These are the same age and the same direction and shape as the Carolina Bays and are from the same period.

Interestingly enough there are the ruins of an ancient city in Point Barrow above the present Arctic Circle. Are our dates correct for these ruins or are they much older than we know?

Ten miles north of Jasper in Boone County in Arkansas the fossilized remains of Paleozoic palm trees, tapirs, saber-toothed tigers, rhinoceroses and other tropical biology have been found in the large Willcockson Fossil Beds nearby. They were not supposed to be this far north and seem to indicate a sudden and drastic earth movement in the remote past. Which Earth movement though? Marble Falls is in Boone County in Arkansas.

There is a mass of traveled rocks at Alpine, twenty miles east of San Diego, similar to Point Loma, that cover an area of thirty acres at an altitude of 1,500 feet above Sea level. These must have been pushed here into this one particular place by severe and sudden oceanic flooding. The area is not glacial. Alpine is in San Diego County in California.

On the tip of a high bluff that juts down from the north to close off San Diego Bay from the Pacific there are hundreds of lava and metamorphic traveled boulders, which are exposed at low tide. Scores of others are eroding on top of the point with one of the largest, at fifty tons, lying 300 feet from the lighthouse at the South tip. The nearest source for the rocks is North Coronado Island, eighteen miles southwest in the Pacific Ocean. These are not the result of glaciation as this is not a glaciated area. The highest rocks are at a level of 350 feet above the highest tide. This is indicative of severe and sudden oceanic flooding. This was at Point Loma in San Diego County in California.

Fossils from Lake Florissant in Teller County in Colorado which was a glacial lake were excavated from volcanic ash formed during the decline of the Wisconsin Ice Cap.

In northern Florida marine deposits from the last Ice Age have been found at altitudes of at least 240 feet.

Off the Florida Keys on the American continental shelf hydrographic surveys made by the U. S. Geodetic survey have revealed 400-foot indentations along a 500-foot bottom presumed to have been freshwater lakes in areas that subsided. Are they lakes or are they meteorite craters similar to those found in the Carolinas?

Massive piles of mastodon and saber-toothed tiger bones were found in Florida from the period of the Cataclysm.

More evidence of killer tsunamis was found in Lake Okeechobee in Florida. There are extensive fossil beds in a 40-foot layer of marl containing diverse deposits of Pleistocene marine fauna interspersed at higher levels with dismembered bones of larger creatures, presumably terrestrial mammals. The mollusks and other marine groups though only belong in groups found in remote areas of the Pacific Ocean. The Pacific fauna might have been deposited when there was no Isthmus of Panama to stop them as if a great flood went over it and then through what is now the Gulf of Mexico to get to central Florida.

South of Tampa in Hillsborough County in Florida one of the richest boneyards in America has been found. The bones of more than seventy species have been found there. About eighty per cent of the bones belong to plains animals such as camels, horses, mammoths and others but also bears, wolves and large cats as well as a bird with a thirty foot wingspan have been found. Mixed in with all of the land animals remains are sharks teeth, turtle shells and the bones of both freshwater and saltwater fish all jumbled together. All of the bones are smashed up as if by one very big catastrophe or a huge tsunami.

The Brushy Creek Impact Crater in St Helena Parish in Louisiana is only eleven thousand years old and is 1.9 kilometres across. Shocked quartz and intensely fractured quartz have been found here. The site is 9.3 kilometres southwest of Greensburg.

Near Cumberland in Maryland in 1912 workmen who were cutting a way for a railroad discovered a cavern filled with the shattered remains of numerous animals from different climatic zones. There are crocodilians, tapirs, wolverines, lemmings, shrews, minks, squirrels, and others such as coyote, badger and a puma-like cat in the cavern as well as numerous other animals, all of whom appear to have been swept into the cavern by a massive onrush of water. They all died at the same time. Now we have had tsunamis from Florida to Maryland. More smashed and mixed animals?

In bogs covering glacial deposits from the last Ice Age the skeletons of two whales were discovered in Michigan. Was there a sea in this area after the Ice Age or was this the result of a massive tidal wave? The altitude of the hills is five hundred feet and it is hundreds of miles inland.

Fragments of well-preserved tree trunks have been exhumed from wells in Ann Arbor in Washtenaw County in Michigan. What sort of land drifts covered these trees up?

As the great Wisconsin ice sheets melted they created huge but temporary lakes, which filled up with amazing speed, drowning everything in their path then draining away in a few hundred years. Lake Agassiz or Agassiz Pool was the largest glacial lake in the Americas and once occupied an area of 110,000 square miles covering large parts of what are now Manitoba, Ontario, and Saskatchewan in Canada and North Dakota and Minnesota in the United States. It lasted for less than a Millennium.

Five kilometers north of Pelican Rapids in Otter Tail County in Minnesota road excavators in 1931 found an unusual human skeleton three metres below ground in lakebed clay from Lake Pelican which was a small glacial lake. The skeleton is called "Minnesota Minnie" and is of a fifteen year old female that was a mixture of Mongoloid and an early white race. "Minnesota Minnie" was wearing a marine shell pendant and there was a flaking tool made from elk antler in association with her. There is a possibility that she drowned. Was this when the glaciers were rapidly melting? Who were the early white race that was in her ancestry as well as the Mongoloid race? We have met this strange white race as well as African races before in North America.

In the United States Ice Age marine features are present along the gulf coast east of the Mississippi River, in some places at altitudes of two hundred feet. Once again two hundred feet up. Was this the extent of the post glacial tsunamis? Or were thse tsunamis just some of them?

Professor Robert. W. Gilder of Omaha, Nebraska, found artifacts such as jugs, carvings, charred sticks, clay pipes, fish-hooks, shell ornaments and elk horn combs in the famous Buffalo Wallows. Gilder postulates that the wallows were originally the entrances to caves in which he has found the artifacts. These caves have subsided and been used later by the buffalos. These artifacts and caves date from before the last Glacial Period or the late Tertiary. These caves were filled with earth and torn up by the waves of the last Cataclysm.

The fossilized remains of rhinoceroses, hippopotamuses and zebras all in swimming positions along with many other animals were found in Nebraska. What were they swimming in or from? Were they trying to escape the glacial meltwater?

An oceanographic survey conducted by Bell Telephone Laboratories stated that a one hundred and eighty kilometre wide submerged strip of land found off the New Jersey coast was dry land between five to fifteen thousand years ago or 15,000 BC. Fossils of zooplankton foraminifera that thrive in freshwater lagoons were found there. With all of the other reports of the continental shelf of the United States being submerged around 9,000 BC then we can safely place this report here as well.

There are very fresh volcanic lava flows in the San Jose River Valley in New Mexico to validate Indian legends from the same area of a river of fire. These lava flows date from around 9,000 BC.

Fossils from the end of the last Ice Age found in the John Day River Basin in Oregon were excavated from volcanic ash.

In Texas well to the south of the furthest extent of the Wisconsin Glaciation the remains of Ice Age land mammals are found in marine deposits.

The skeleton of a whale from the last Ice Age has been found in Vermont more than five hundred feet above the sea. Was it travelling along on land?

In North America the last Ice Age is called the Wisconsin Glaciation. It started around 115,000 years ago. There were various advances and retreats of the ice after that with the fastest rate of accumulation taking place between 60,000 and 70,000 years ago and 17,000 years ago. This culminated in the Tazewell Advance. The maximum extent reached was in 15,000 BC. By 13,000 BC millions of square miles of ice had suddenly melted. Around 9,000 BC the ice expanded again but then retreated. By 8,000 BC the Wisconsin Glaciation had withdrawn completely.

In Two Creeks in Manitowoc County in Wisconsin the last advance of the glacial ice crushed a pine forest and snapped off the tree trunks and left them with their tops pointing southwards in the direction of the advance. Then the ice stopped prior to retreating suddenly. This was approximately 11,000 years ago. It was also reported that as well a spruce forest from Two Creeks, Wisconsin, was crushed by advancing glaciers about 11,000 years ago. Advancing glaciers though raze the surface of the earth like bulldozers and crush trees. This is evidence of another massive tsunami.

Berlitz, 1969, mentions whole *mastodons*, *toxodons*, giant sloths and other animals that were found quick-frozen among the mountain glaciers in Venezuela.

Advancing ice masses about 10,800 years ago were reported to have uprooted birch trees in Germany. Wouldn't ice masses have crushed them? Would it not be more like massive tsunamis pushing and uprooting the trees? Is this the same as in Wisconsin and so many other places including Siberia and Alaska?

Nevali Cori in the Kantara Valley in Eastern Anatolia in Turkey was settled from 8,800 BC to 7,600 BC. It is now flooded by the Ataturk Dam on the Euphrates River. The site was excavated in 1983 and drowned by 1991. Here there was an enormous sculptured monolith standing in the middle of a sunken temple with walls composed of dry pack stone interspersed by a series of upright pillars, similar to Tiahuanaco in Bolivia. The monolith was precisely rectangular and set into a perfectly smooth stone floor, which was composed of a lime-based mortar known as terrazzo and was unique to Neolithic Sites here. On the two metre tall pillar or monolith which had obviously lost its uppermost section there were two low relief arms bent upwards so that they formed an horizontal v-shape. These terminated on the front narrow face of the monolith in stylized hands each with five fingers of equal length, like flippers of a seal, a whale or a dolphin. Above the hands were two long rectangular strips that ran from behind the visible break at the top of the stone to about halfway down its length giving the impression of extremely long hair hanging over the shoulders of a human form. The society that first settled here in 8,400 BC already knew of the basic principles of agriculture. Of the 22 rectangular buildings in a grid shaped layout here only one was apparently used as a domestic dwelling. Most of the site was used for storage alone. One was a workshop for the making of flint implements and another had possible cultic use in the form of buried skulls. The primary function of the community was as a religious centre. There were dry walls of stone here inside an earthen bank, which rose to 2.8 metres high and half a metre thick. The oldest wall was four metres long. The walls had been covered by limestone mortar showing traces of grey-white plaster bearing evidence of black and red paint and murals. There was also evidence of human sacrifice. The body of a female was found inside one of the structures with flint chips embedded in her head and neck evidence of being killed with flint projectiles.

Deep core samples taken from the Greenland Ice Cap in 1989 revealed that around 8,700 BC the last cold period of the Ice Age came to an abrupt halt. The ice retreated so quickly that major climate changes occurred within twenty years and a major temperature rise of seven degrees Centigrade occurred in fifty years. Later core samples indicated that the most significant melting and collapse might have occurred in one to three years. The turbulent sea would have literally rushed over hundreds of miles of plains and forests engulfing human settlements in its path. What is it with these fifty year long sudden temperature rises?

Around the middle of the ninth millennium BC geological evidence shows that there was a mini-Ice Age in Turkey.

In Death Valley in Inyo County in California there is ample fossil and skeletal evidence to indicate that the area was once a tropical paradise with ample lakes and forests for habitation. It is now an area of total desolation. There were still massive lakes in the area up until 7,000 BC with the last lake drying up three thousand years ago. The lakes originally covered the valley floor which is several hundred feet below below sea level and started drying up around 8,500 BC.

The Parry Sound Impact Crater in Ontario in Canada is ten thousand one hundred and fifty years old and is 2.5 kilometres in diameter. How come every other time the geological periods have massive dating variations yet this time it is down to fifty years! Is this correct?

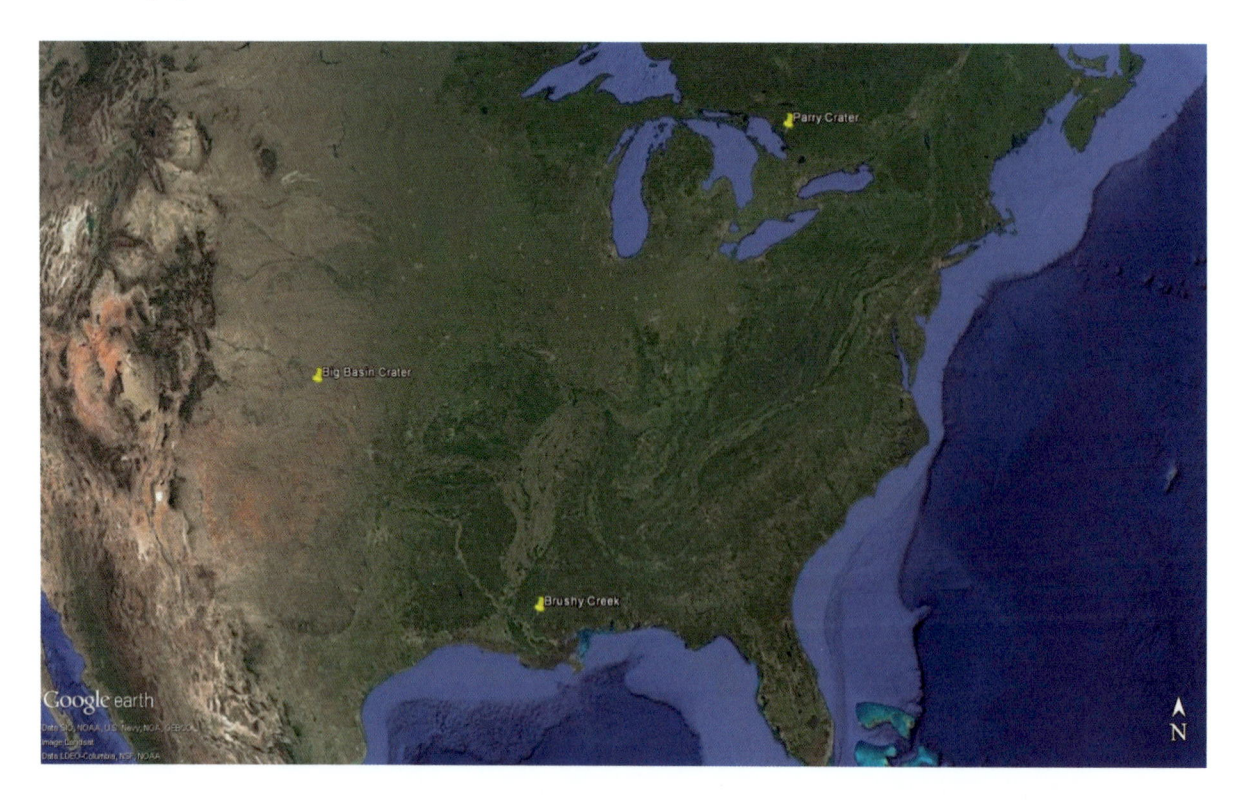

What meteoric events happened in the Ninth Millenium? The tally is the Morro de Cuero Crater in Argentina, the Curtis Lake Crater near Palmer in

Alaska, the Brushy Creek Impact Crater in St Helena Parish in Louisiana and the Parry Sound Impact Crater in Ontario in Canada. Not many really when compared to other millenia. Not as many as in the Tenth Millenium but still significant.

8th Millenium BC.

Now the party really starts. And you did not even know that there had even been a party in the first place.

The period around 8,000 BC is far more complex and far more extraordinary as the period around 9,000 BC and 10,000 BC. Allowing for geological variance, are they all the same period including a lot of different opinions set around the same period of dates? Do we in fact have to compress the data from both of these periods to get a real view of the phenomenal events that happened in this time? Are all of the reports from 10,000 BC to 8,000 BC all contemporary to each other?

Kelly and Dachille state that the extinction of mastodons, like mammoths, occurred so fast 10,000 years ago around 8,000 BC, that they seemed to have been rapidly frozen and that this was a result of a shift in the Earth's crust following an impact with a celestial body. Indian elephants did not become extinct because they were too far from the new Pole, as in North Pole. It was not a case of survival of the fittest but survival of the luckiest. The sudden freezing of the animals could only have come about because of the sudden changing of the latitude of the region where they lived. The frozen creatures from formerly warmer climes are only found in one half of the frozen Arctic Hemisphere, which suggests how the Pole Shift happened, and possibly its point of impact. These animals are now found where the ground is frozen for several feet but when they fell the ground could not have been frozen. Rather they fell and were then covered with water or mud that froze and has never thawed since. In some places the debris in which the animals are buried is thirty metres deep. The recovery of perfectly preserved grass and flowers in the stomachs of the creatures indicate that the Cataclysmic event occurred in the summertime. Does this sound like I am repeating myself? Maybe we do need to compress the alleged data from both of these periods? When you see the views of earth from above you will be amazed.

Some sources state that the north-south Axis of the Earth shifted 10,000 years ago. Did an impacting comet or comet particle cause the tilt or flip?

Is the earth tilt a self-regulating mechanism like a wobble in a gyroscope or is it due to a sudden impact from elsewhere like a billiard ball being hit with a cue or another ball? 180 degrees would have seen north become east or west and vice-versa with south. The Poles would have moved considerably.

The further study of fossil magnetism indicates that there have been at least 170 planet tilting cataclysms in the last 80 million years. There were 170 Cataclysms at least? In only the last eighty million years? Movements of the Magnetic Poles have occurred in total 276 times that we know of, from only a few degrees to total 360 degree shifts since the formation of the earth.

There was even a planet shift for the accepted end of the Dinosaurs sixty-five million years ago when a cosmic body was theorized to have hit the Gulf of Mexico forming the Chicxulub Crater? This massive impact would have caused a degree of planet tilt. Looking back on it now, Chicxulub was small compared to the Cuba Crater next door from the same period. You never hear of the Cuba Crater though. Funny that.

From 8,000 BC to 4,000 BC there was a period of global warming called the "Neolithic Climate Optimum" when world temperatures rose on average accompanied by more rain.

The Mullion culture suddenly appeared on the coast of Algeria 10,000 years ago leaving the largest skeletal population in the world. The Mullions also possessed the largest cranial capacity of any population with approximately 2,000 cubic centimetres compared to modern man's 1,400 cubic centimetres. The bodies were short and spindly with large heads. The Mullion only inhabited the site briefly and their population consisted mostly of women and children who worked with tool types and domesticated animals never before seen. Where they came from is unknown but there were several population movements during this period such as the then contemporary and also suddenly arriving Azilian migrations into what is now Spain. Had these two groups of people, since we have no record of their movement, actually come from the west, from the now vanished Atlantean continent or archipelago? Were they a flood of Neolithic refugees? From a cosmic body caused disaster? Or were they a flood of refugees from the suddenly drowned Mediterranean Valley? Or both?

Though the Wurm Glaciation suddenly ended in 8,000 BC in Europe glaciation only started in the Arctic in 4,000 BC.

When Charles Darwin visited La Plata in Argentina he found the tooth of a horse embedded with the remains of a *megatherium*, a *mastodon*, a *toxodon*, and other extinct animals which all co-existed at a very late geological time. Darwin was filled with astonishment seeing that the horse since its introduction by the Spanish in the sixteenth century had run wild over the whole country at an unparalled rate. What could have so recently, geologically-speaking, killed the former horse under conditions of life that now appeared so favourable?

The final and major extinctions of the last Ice Age were over seven thousand years from 15,000 BC to 8,000 BC. This is when the original American horse allegedly became extinct.

The Rio Cuarto Crater Field in Cordoba Province in Argentina is ten thousand years old. There are ten craters in total. One is named the "Drop" and is two hundred metres wide and six hundred metres long. Two other craters are the "Eastern Twin" and the "Western Twin" which are both seven hundred metres wide and 3.5 kilometres long and are five kilometres to the northeast. The "Northern Basin" crater is slightly over one kilometre wide and 4.75 kilometres long. The long axes of all of the craters point to the northeast and the elliptical craters were formed by very low angle impacts. Impacts over forty-five degrees usually form circular craters whereas elliptical craters are created by low trajectory impacts. The Morro de Cuerra Crater near Tunuyan in Mendoza Province in Argentina is ten thousand years old and 600 metres in diameter. This crater is five hundred and fifteen kilometres west from the Rio Cuarto Crater Field.

Near El Carmen in Jujuy Province in Argentina a cave was found that contained paintings similar to those in Northern Amazonas in Brazil and depicting horses and horsemen as well as double headed axes and bulls heads amongst other things. The horse is supposed to have been extinct in the Americas since 8,000 BC and to never have been ridden in that period. These cave paintings are much older and show precisely the opposite. There were no bulls in the Americas at any time and these are not bison. The double-headed axe was a prominent symbol in ancient times especially in Europe and the Mediterranean. This symbol is even on one of the trilithons at Stonehenge.

Around 8,000 BC sea temperatures in the Northern Atlantic Ocean suddenly dropped.

The Mid-Atlantic Ridge is a mountainous ridge up to two miles high and hundreds of miles wide running in an S-curve down the Atlantic midway between the Americas and Africa and Europe following the contours of those continents. Islands such as the Azores, Ascension, and Tristan da Cunha mark its course. The mid-Atlantic ridge was above sea level according to samples taken from the Sierra Leone Rise. This was 10,000 years ago and the samples were found at a depth of twelve thousand feet. Core samples contained freshwater diatoms, a life form seen in freshwater lakes. There must have been freshwater lakes here in that period and these freshwater lakes had dropped twelve thousand feet all of a sudden.

There are two types of foraminifera which are little organisms that live in the water. The two principal species are *globorotalia menardii* and *globorotalia truncatulinoides*. The first is distinguished by a shell spiraling left wise and lives in warm water. The other spirals right wise and lives in cold water. The warm type does not appear anywhere above the line stretching from the Azores to the Canaries. The coldwater foraminifera are present in the northeastern quadrant of the Atlantic. The warm type inhabits the middle Atlantic zone from West Africa to Central America. Yet in the equatorial Atlantic the cold type show up again. It looks as if the warm species of foraminifera tore through some barrier and headed in an easterly direction. Based on foraminifera distribution the Lamont Geological Observatory discovered that a sudden warming of surface ocean waters occurred in the Atlantic about 10,000 years ago. The transformation of the cold type foraminifera to the warm type did not last more than one hundred years.

The bed of the Mid-Atlantic Ridge sank about ten thousand feet where it was previously above sealevel.

Around ten thousand years ago the world's largest volcanic eruption occurred on a large landmass on the Mid-Atlantic Ridge in the Atlantic Ocean. Projectiles from this eruption were thrown as far as the Gulf of Mexico and the Southeastern United States. This force was reckoned to be the equivalent of 15,000 hydrogen bombs going off simultaneously and this was probably caused by an asteroid, a comet or an invisible energy body during one of the most amazing meteor showers in the Earth's history. The volcanic eruption was caused by the impact as the crust of the earth split open and buckled. In Labrador in Canada there was another impact that sent projectiles hurtling for thousands of miles. All over the eastern seaboard of the United States thousands of meteors left deep elliptical impact craters all heading in the one direction. The force of all of these impacts was enough to send the Earth reeling and teetering as it readjusted its angle of rotation. The Earth moved 28.5 degrees relative to its previous position causing even more disastrous consequences for those unfortunate enough to survive the initial impacts.

A Swedish Oceanographic expedition on board the "Albatross" during 1947-1948 sampled the seabed here and found that perhaps the Mid-Atlantic Ridge in this area was actually above sea level 10,000 years ago. The cores that were taken were from a depth of 12,000 feet and carried samples of freshwater diatoms, a life form only seen in freshwater lakes. Obviously freshwater had existed here before. At the point where the most diatoms were found the Sea was now 8,000 feet deep!

Sir William Dawson believed that extensive submergences took place at the end of the last Ice Age, not more than ten thousand years ago around 8,000 BC.

Pilots and observers flying between Senegal in West Africa and Brazil have reported seeing submerged walls of stone and ruins at approximately twenty degrees west and six degrees north near the Sierra Leone Rise.

According to Rene Malaise, a Swedish scientist and geologist, the present terrain of the sea bottom here has more to do with above water weathering processes than by the actions of undersea currents.

Incidentally on the Piri Reis Map one of two large islands shown in the Atlantic Ocean was located where the Sierra Leone Rise is now.

In this period the vast inland seas of Australia suddenly dried up leaving huge areas of parched desert and glistening salt lakes. Was a planetary shift responsible for changing the weather and climatic patterns over the continent?

The remains of *diprotodons*, the largest marsupial that ever existed, were discovered in the Wellington Caves in New South Wales in Australia by Sir Richard Owen in 1838. The *diprotodon* was a giant relative of the wombat and twice the size of a rhinoceros. The *diprotodon* suddenly became extinct around 10,000 years ago. The cave is full of the skeletal remains of marsupials with heads, jawbones, teeth, ribs and femurs all concreted together. Had a giant wave washed them into the cave system? Incidentally the remains of people were found in the Wellington Caves that lived here at the end of the Tertiary Period. The remains are over one million years old! Humans were not supposed to be in Australia one million years ago! Aborigines were only supposed to have arrived in Australia at the most fifty thousand years ago. This is under debate though. We have already seen possible earlier arrivals into Australia.

At Lake Callabonna in South Australia in 1892 and only a few feet from the dried-up lake surface a whole graveyard of bones of the *diprotodon* were discovered. The *diprotodons* suddenly became extinct 10,000 years ago. Was it mass suicide? Seems unlikely if the *diprotodons* were that organized or self aware.

The jumbled bones of a variety of extinct creatures were found in the Naracoorte Cave system in South Australia. They appeared to have rushed into the cave system at the time of their extinction around 10,000 years ago around 8,000 BC. There was massive and sudden extinction of animals here. Were there also massive tidal waves rushing over the landscape in this period?

The Joey Crater is in the Southern Ocean south of Australia between the Great Australian Bight and Bass Strait. It is situated on the continental shelf of Australia. The crater is ten thousand years old and four kilometres across. South of the Joey Crater is the Kangaroo Crater that was created around the same time. Naracoorte is north of here.

The Kangaroo Crater is in the Southern Ocean south of Australia between the Great Australian Bight and Bass Strait. It is situated on the continental shelf of Australia. The crater is ten thousand years old and five kilometres across. These are two very likely contenders for the creators of tidal waves over parts of South Australia. This area of the continental shelf would have been above sea level in this period as were most of the continental shelves.

The Flinders Impact Crater was formed ten thousand years ago in Bass Strait between Tasmania and the Australian mainland. The crater is ten kilometres across. Other sources state that the impact occurred one hundred thousand years ago. The area where Bass Strait is now was above sea level until this period ten thousand years ago when the ocean crashed over it.

The Hickman crater is ten thousand years old, some say one hundred thousand years, there's that geologic leeway again, and two hundred and sixty metres wide and thirty metres deep. The Hickman crater is thirty-five kilometres north of Newman in Western Australia.

Tektites found in Victoria have been dated to 8,000 BC. These were produced by an impact southeast of Tasmania. Was this particular Tektite shower connected with the impact of the comet, or meteor, or asteroid, in the Tasman Sea to Victoria's east? Or were the tektites produced by Joey 1 and

Joey 2 in the Southern Ocean? In the same period meteoric storms also hit the Tasman Sea off Eastern Australia as well as the China Sea off Vietnam. Then again there might be still undiscovered impactoid craters in the Tasman Sea as well.

These are the impact craters in Australia from 8,000 BC. The Kangaroo and Joey craters are together south of South Australia, the Flinders Crater is in Bass Strait between Tasmania and Victoria which was also above sea level in this period and on the mainland in Western Australia is the Hickman Crater. To the southeast of Australia is the Kukla Crater in the Southern Ocean.

Around eight thousand BC up to two thousand BC the broad plateau of the Bahamas was flooded leaving only a shallow sea and assorted islands. Rising world sea levels would have caused this as the glaciers of North America suddenly melted.

The eastern part of the Caribbean Sea bordered by the arc of the Lesser Antilles according to depth charts indicates a deep sea hole right in the middle of the Antilles Arc which surrounds it like the rim of a half submerged giant crater. Kelso de Montigny thinks that a giant asteroid struck this area 10,000 years ago around 8,000 BC. The hole is smaller and shallower than the two near Puerto Rico from the same time and may have been caused by a fragment of the larger impact.

There is a network of river-channels on the seafloor between Sumatra and Java in Indonesia and Borneo. This was called Sundaland and many of the drowned rivers are extensions of present ones on the islands. There are also drowned offshore peat deposits from this period as well as undisturbed tree stumps. Peat is only created above water level and not the sea. There was a crustal subsidence in this area and the invasion by the sea of vast tracts of land covering a vast area from Indonesia to Borneo around 8,000 BC. The date of the flooding of Sunda Land is already quite flexible. Once again we might have to compress our data periods together. With the gaps that we have seen so far one or two thousand years is nothing.

The American West coast rose from the sea to a level of 2,000 metres around 10,000 years ago.

Along the Arctic coast of Canada there is a marine deposit containing walrus, seals and at least five genera of whales overlying the seaboard. The same is along the northeastern states of Canada as well.

In many areas of the Pacific North West coast of North America Ice Age marine deposits have been found more than two hundred miles inland. Were these from massive tidal waves?

Along the coast of Newfoundland and New England there are numerous stumps of trees in the water that indicate that massive forested areas became submerged at the end of the Ice Age.

The Merewether Crater in Labrador in Canada is ten thousand years old and two hundred and thirteen metres across.

The Hudson Bay Crater has a nearly perfect circular arc that covers more than one hundred and sixty degrees of a circle with a diameter of four hundred and fifty kilometres. The Belcher Islands are where the central uplift would be. The arc is called the Nastapoka Arc. A direct hit by a comet or massive asteroid here would have ended the last Ice Age by burning and impacting through what was then the ice cap and causing it to rapidly disintegrate and melt. Watermelt under the mile thick ice masses would have sped up the breaking down of the ice and caused massive water rushes or landbased tsunamis. Yes, we have met this before as well but you know my reasoning.The Hudson Strait crater was formed ten thousand years ago. It is three hundred and twenty kilometres in diameter and was possibly one of the meteorite or cometary impacts that ended the last Ice Age. Several researchers theorize that the southeast cormer of Hudson Bay was also an astrobleme that was created around the same time as the Hudson Strait Crater.

These are enormous impactoid impacts and would have created immense damage to what was then the North Polar ice mass over Hudson Bay.

Ungava Bay, which separates Baffin Island from Nunavik in far northern Quebec, is an oval-shaped bay about two hundred and sixty kilometres wide by three hundred and twenty kilometres in length. The crater here is believed to be two hundred and twenty-five kilometres in diameter.

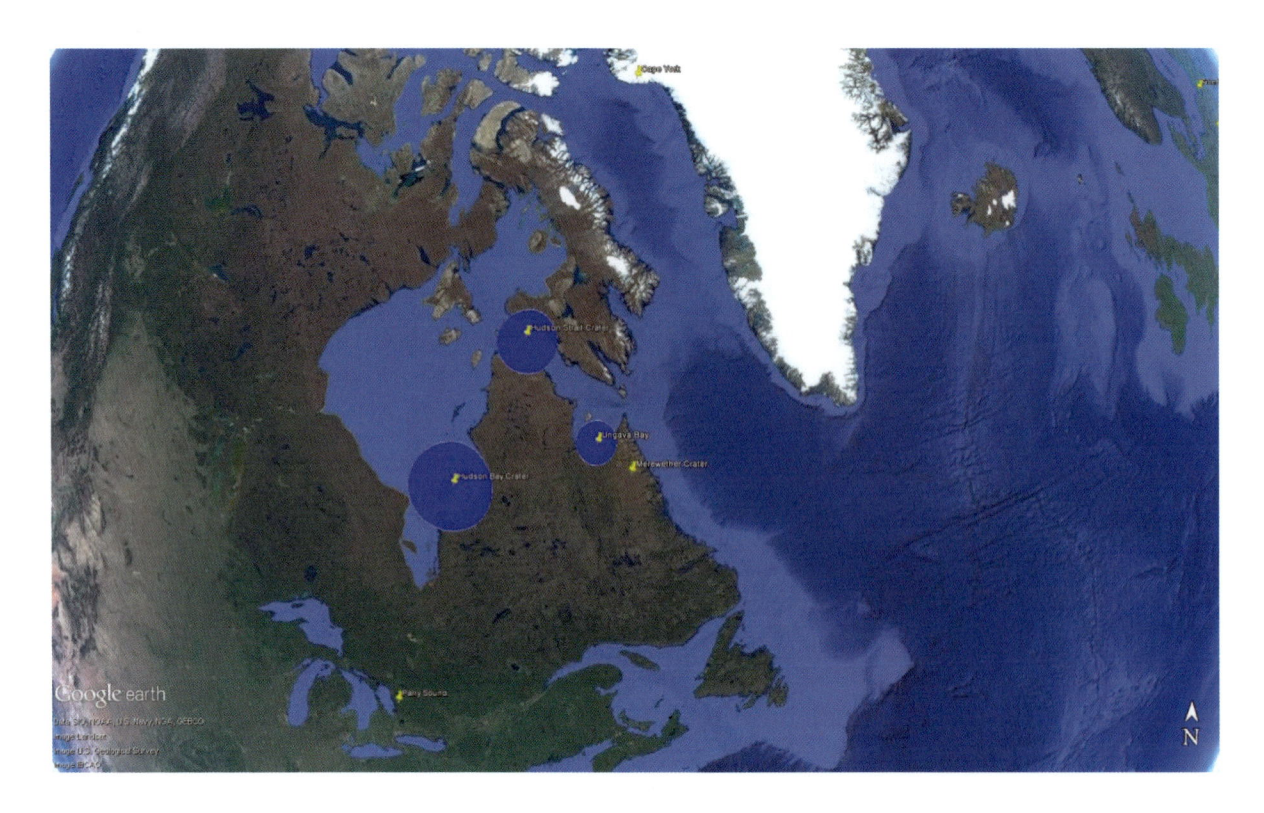

This is a view of 8,000 BC impact events in Canada and Greenland. Hudson Strait, Ungava Bay and Merewether Craters are almost on a great curve of the earth. The Cape York crater, Hudson Strait and Hudson Bay craters also form a line. Or was it a very wide meteor stream over Canada and Greenland? Were these craters all formed at the same time?

The Atlantic islands of Palma, Fuertoventura and Gomera are parts of an older mountain range consisting of diabase, an eruptive volcanic rock. On Grand Canary and Palma an upheaval of from 600 to 1,000 feet can be demonstrated and in Madeira up to 1,400 feet. Professor de Lapparent favoured the existence of a coastline or an island chain during the Miocene Era connecting the West Indies with southern Europe. The end of the Pleiocene and the whole of the Pleistocene Period were distinguished by a series of subsidences, which resulted in finally opening up the northern depression of the Atlantic Ocean.

There exists a small blind species of shrimp in one black tidewater pool in a cavern on Lanzarote in the Canary Islands. The shrimp is called *munidopsis polymorpha* and has residual eyes but has lost its sight and is closely related to *munidopsis tridentata*, only different because it can see. Was *polymorpha* trapped in the cavern due to seismic activity?

Between 8,000 BC and 6,000 BC analysis of ancient lake sediments from Lake Chad show high beach levels which are indicative of a wetter environment. Standing water here formed lakes that lasted considerable periods. The wet period affected all of North Africa and East Africa. Grasslands around these lakes which had attracted migrating sheep herders from the Middle East lasted until 2,500 BC.

At Palli Aike, Palliache, on the Straits of Magellan in the extreme south of Chile there are the remains of humans who inhabited the area from 12,000 BC to 8,000 BC. Five cremated human remains were found here. The remains were found by Junius Bird in 1936 inside a lava tube. The remains of the giant sloth and the American horse were also found here.

The South American West coast rose from the sea to a level of 2,000 metres around 10,000 years ago or 8,000 BC.

All across what is now China there was a sudden dropping of lake levels and a rapid increase in aridity. Climatic conditions and plant life were affected as there was a sudden northern displacement of a now greatly enhanced monsoon system which did not move back to its present position further south until 4,000 BC.

The Gobi Desert was originally a Sea. The Gobi Desert is called the Sham Ho in Chinese, the "Dried Up Sea". How did the Chinese in ancient times know that the Gobi Desert was once a sea? In the Gobi Desert there are huge masses of rock that have been fused together under intense heat and the rocks have been vitrified by huge heat sources coming from above. A stratigraphic layer of fused green glass has been found here that could only have been caused by a nuclear explosion. This stratified green glass has been found worldwide in many areas that are incidentally all deserts now! Are these the results of impacts of our invisible heat sources that we have already met that occurred in conjunction with the cosmic bombardments? Some sources state that this occurred ten thousand years ago around 8,000 BC. Archeologists and geologists dates are as we know quite variable.

The South American West Coast rose from the Sea to a level of 2,000 metres around 10,000 years ago or 8,000 BC.

A vast number of suddenly deceased animals from this period have been found at Santa Fe de Bogota in Bogota Province in Columbia. Atop the Andean highlands is the graveyard of the mastodons. Bones were found petrified and the animals were killed in a great earth upheaval. At the time of their deaths the animals were feeding in a lush, warm jungle near the shores of the Pacific Ocean. Suddenly the mountains were raised and the great beasts died in the rarified and cold air. There is an imposing plateau here called the Giant's Field. It is full of huge fossilized bones. Von Humboldt saw the remains of a mastodon, a creature nearly as big as a mammoth but with short stumpy tusks and a trunk almost as long as its body. It preferred marshy areas with plenty of vegetation. It could not have survived at the present altitude of 6,500 feet. The petrification of the bones could only have been done by sea-salt and the area must have been raised to its present altitude after the mastodon had died. How did the saltwater get up the mountain? The elephants here appeared to have died as if they were struck down during a massive stampede. Was the mother of all tsunamis chasing them as the plains suddenly became mountains? What had struck them down?

When the American submarine "Nautilus" made her voyage around the world she called attention to the presence of an exceedingly lofty and still unidentified underwater peak close to Easter Island.

Professor H. W. Menard of the University of California speaks of an exceedingly important fracture zone in the neighborhood of Easter Island, a zone parallel to that of the Marquesas Archipelago. Professor Menard also speaks of the discovery of an immense bank or ridge of sediment.

According to tradition the islet of Sala-y-gomes, three hundred miles east of Easter Island, was called Motu Motiro Hiva, Small Island near Hiva. Hiva was the fabled lost continent of the Easter Islanders. Legends state that Easter Island was once a much larger country but because of the sins of its people Uoke tipped it up and broke it with his crowbar. The island was once part of a large archipelago that is now beneath the sea. Can a descending cosmic body look like a crowbar? Can it look like a crowbar when you see its' sideways view with a long straight tail?

The South American west coast of Ecuador rose from the sea to a level of 2,000 metres around 10,000 years ago. This is the same reported height as the West Coast of Canada.

The El Fayum or El Fayoum depression is a ten thousand year old impact structure south-west of the Pyramids of Giza in Egypt. The depression was created ten thousand years ago and is one hundred kilometres in diameter. Just a note but if there were early memories of this comet or asteroid coming from the direction of certain constellations, could the siting of the pyramids of Giza be a massive commemorative structure for what would have been an horrendous event! Did the impactoids come from the direction of the Orion Constellation as Robert Bauval believed them to be positioned so as to be a map of Orion? Some archeologists state that the Sphinx of Giza represents the vernal equinox of Leo which was supposed to have occurred around this period. Or was it just a massive coincidence?

Nabta Playa is sixty-five miles west of Abu Simbel in Egypt and is in the Nubian Desert. Around ten thousand years ago rainy conditions had moved out across Egypt and numerous seasonal lakes and playas were formed. One of these lakes and one of the largest in southern Egypt was at Nabta. Pastoralists entered this area during the summer and then left in winter but would return the next summer. The remains of hundreds of Stone Age camps have been found here in a large ancient lake basin. There were grave mounds with offerings of butchered cattle, goats and sheep as well as groups of megaliths and alignments of upright stones. These megaliths were discovered only in 1972 by archeologists Fred Wendorf and Romauld Schild. There are six groups of stones extending across the basin which contain a total of twenty-four megaliths. Like the spokes of a wheel they are described as radiating out from a unique and complex structure and span 2,500 metres in a north-south direction. At the northern end is a stone circle and ten burial mounds and the stone circle sits on a small hill and it has been decided that it is a calendar and it appears to have been last used around 4,000 BC. This is a very early ceremonial centre marking the beginning of complex societies in Africa. To the south of the of the basin is a low elongated hill with a six hundred metre alignment of upright sandstone megaliths weighing several tons. There are numerous other lines and sublines of megaliths as well with several astronomical assertions being placed in regard to them. One 250 metre line in a double alignment aims at where the brightest stars in Orion's Belt would have been between 6,170 and 5,800 years ago. A second line of stones points towards where Sirius or Canis Major would have been 6,800 years ago. These were the two most important constellations in the Egyptian pantheon when that civilization rose up much later. Around 7,000 BC people who had been nomads had now settled here and built deep wells and organized villages of small buildings set in straight lines. Is this also a memorial to the impact that created the Fayoum Depression?

Some authorities state that Britain was attached to Europe until 8,000 BC when the connecting land sank forming the North Sea and the English Channel. There is agreement that the English Channel was formed between 7,500 BC and 5,500 BC. Once again, pick an age in this same period.

During the greatest advances of the Pleistocene ice England and Ireland were joined to Europe and most of the North Sea was a low plain over which the Thames, Rhine and other rivers sluggishly wound their ways. The plain sank beneath the sea between 25,000 and 10,000 years ago, 23,000 BC to 8,000 BC. Today fishermen dredge up Stone Age tools and mammoth teeth from the sea bottom. Fishermen also do this off the east coast of North America. The meltwater from the melting glaciers would have raised world sea levels immensely.

Wide plains extended from the present day English Channel into the Atlantic.

The English Channel separating England from Europe was formed. Geologists state that less than nine thousand years ago oak trees were growing where the water is now sixty metres deep.

Near Cromer at Mundesley in Norfolk the remains of submerged forests have been found in the North Sea. The remains of upright stumps that are interlocked together have been found that appear to have been uprooted and then moved en masse. The remains of sixty species of mammals besides birds, frogs and snakes have been found inside these stump beds. Most of these animals are now extinct and included temperate as well as northern and tropical latitudes. There were northern species like glutton and musk-ox as well as saber-toothed tigers, huge bears, mammoth, elephant, hippopotamus, rhinoceros and horses amongst others.

The Tsoorikmae Impact Crater in Estonia is forty metres in diameter and ten thousand years old. It is near Tsyyrikmyagi in Estonia.

The Cabrerolles Crater is two hundred and seventeen metres across and was created ten thousand years ago. There are five other meteoric craters in total here that are from two hundred metres wide down to fifty-five metres wide. This was a minor impact swarm. The Cabrerolles Crater is near Faugeres in Cabreron in Clermont-l'Herault in Herault in France.

There are repetitive marks in a cave at Rochebertier in France that may be picture writing, a tally or even an alphabet. The marks though are 8,000 to 10,000 years old from 6,000 BC to 8,000 BC. Rochebertier is near Mazerolles in the Charente in France. Isn't this the same period as the marks found at Mas d'Azil, which is near Foiz? Also found here is a twelve thousand year old, 10,000 BC, reindeer bone that has markings on it that appear to be more than just decoration. They appear to be letters or some form of writing. The writing is similar to the Tartessian script of Iberian Spain that is supposed to be six thousand years later.

More deceased animals were found in the Lazaretto Grottoes near Nice in Provence-Alpes-Cote-d'Azur in France. At the start of the Twentieth Century a dynamite explosion brought to light the grottoes and the remains of horses caught in a stampede as well as elephants over ten thousand years old as well as stone tools cut in a very rudimentary fashion. There had been previous hominid remains found at this same site later on that apparently dated back 150,000 years. In May 1964, Francois Octobon discovered the brow of a creature 150,000 years old. The traces of cerebral vessels indicated mental activity far from insignificant. Instruments found nearby included stone axes, knives, rasps and graving tools as well as remains of fires. There were also awls, daggers and extraordinarily well-balanced clubs as well as a small handle of deer bone that was very skillfully split and scapula bones inserted, namely prehistoric razor blades. This was contemporary with *pithecanthropus* on the other side of the Mediterranean in North Africa.

The Cape York Meteorite is around ten thousand years old and left several iron masses. One of these was called Ahnighito, "The Tent", by the Inuit and weighs thirty-one metric tons. Another called "The Woman" weighs three metric tons and there is a third called "The Dog" which weighs four hundred kilogrammes. A fourth fragment called "The Man" was found in 1963 which weighs twenty metric tons. Other fragments have been found from three tons in metric weight down. The Inuit used them as a ready source of iron. These meteorites were found on Saviksoah Island in Savissik in Greenland.

Further developments in research in the Gulf of Mexico show from the study of fossil shells of Foraminifera, a marine life form, that between eleven to twelve thousand years ago, 9,000 to 10,000 BC, there was a substantial increase of twenty per cent in the warmth and a twenty per cent decrease in the salinity of the water. This indicates that there was a profound change in the level and the temperatures of the oceans as glaciers which are composed of freshwater started to melt at this time. There are two types of Foraminifera. The two principal species are *globorotalia menardii* and *globorotalia truncatulinoides*. The first is distinguished by a shell spiraling left wise and lives in warm water. The other spirals right wise and lives in cold water. The warm type does not appear anywhere above the line stretching from the Azores to the Canaries. The coldwater Foraminifera are present in the northeastern quadrant of the Atlantic. The warm type inhabits the middle Atlantic zone from West Africa to Central America. Yet in the equatorial Atlantic the cold type show up again. It looks as if the warm species of Foraminifera tore through some barrier and headed in an easterly direction. Based on Foraminifera distribution the Lamont Geological Observatory discovered that a sudden warming of surface ocean waters occurred in the Atlantic about 10,000 years ago or 8,000 BC. The transformation of the cold type Foraminifera to the warm type did not last more than one hundred years.

The Kukla Crater in the Indian Ocean is ten thousand years old and sixteen kilometres in diameter. Some studies suggest that megatsunamis that hit Africa, India and Southern Australia at this time were caused by this particular impact.

Interesting discoveries were found near Kermanshah in Kurdistan in Iran. Found in Ganj Dara were the remains of mainly sub-adult male and mostly older female goats which indicated selective breeding and culling around 8,000 BC. This killing pattern is indicative of the animal husbandry practice of slaughtering surplus rams as they reach maturity and where the older females are kept for breeding purposes until they are too old to breed. Fired pottery and tiny clay figurines have been found here dating back to the early eighth Millennium, 6,000 BC. These are far in advance of the stone, wood, basketry and plaster typical of this period.

On the shores of the Caspian Sea in the caves of Hotu in Mazandaran grains of domesticated wheat were found that were ten thousand years old. The anthropologist Carleton Coon and the geologist Louis Dupree discovered in some of the caves the remains of three indisputably human creatures that lived here over one hundred thousand years ago.

There is a Kurdistan in Iran as well as Iraq. Now we come to the Iraqi part. Clay tokens used for bartering and trading between different communities were first developed in the foothills of Kurdistan during the eighth millennium BC. These eventually became smaller and more complex and by 3,000 BC had been replaced by sequences of markings on clay cases. Shortly after the first known examples of baked clay tablets bearing ideograms began appearing on the plains of Iraq. Why do I drop in cultural relics of civilization? These cultural relics indicate the presence of intelligent witnesses who witnessed the events and possibly left legends and memories.

Evidence of beaches, particularly around Rathlin Island in Northern Ireland, show that the sea had risen and cut off Ireland from Britain during the Boreal period around 8,000 BC. This was the period of three hundred foot sea levels rises.

Some geologists have indicated that the River Jordan and the alluvial plain around it were created at the end of the last Ice Age around 8,000 BC. This also includes what is now the Dead Sea.

There is an underwater monument at Yonaguni Island off the coast of japan that is believed to be ten thousand years old. Yonaguni is in southwest Japan. Yonaguni is the westernmost point of Japan, closer to Taiwan than Japan itself. Off Arakawa Bana headland just off Isseki Point, translated as Monument Point, in 1987 Kihachiro Aratake discovered an apparently manmade structure carved out of solid rock in complex shapes and patterns that lay with its base on the ocean bed at a depth of 27 metres. It was more than 200 metres long and rose gracefully in a series of pyramidal steps to a summit platform just 5 metres below the surface. Professor Masaaki Kimura states that blocks carved off during the creation of the structure are not found lying in places where they would have naturally fallen. In several small areas of the monument there are completely contrasting features close to each other such as a raised edge two metre deep circular hole, a stepped, cleanly angled geometrical depression and a perfectly narrow straight trench. On the higher surfaces of the structure there are several areas that slope quite steeply toward the South. Deep symmetrical trenches on the northern elevations of these could not be formed by natural means. There is a series of steps rising at regular intervals up the South face of the structure as there is on the northern face. A distinct wall encloses the western side of the structure and it is not natural because it is made of limestone not indigenous to the area. What looks like a ceremonial pathway winds its way around the western and southern faces of the monument.

The Togyz Crater in Kazakhstan is ten thousand years old and twenty kilometres in diameter.

There was another impact in Kazakhstan in this same period. The Chelkar-Aralskaya astrobleme was created ten thousand years ago and is four hundred and twenty kilometres in diameter. It is believed to have been caused by a cometary impact. How many major impacts were there in this period of only one or two thousand years? What of the Aral Crater that was four hundred and twenty kilometres across that was supposed to have occurred in the Eleventh Millenium BC only three thousand years before and in the same area?

The Chelkar-Aralskaya Crater, the Aral Crater of 11,000 BC and the Togyz Crater. Is it coincidence that the two gigantic impacts were in the same area and only three thousand years apart? Were they the same impact even though there is a thirty kilometre size difference as well as different centres? Were they actually created at the same time? Geology quite often contains data with error factors up to fifty million years. I have found impact events occurring in streams when the data is compressed. What then is three thousand years? It is nothing. We have too many apparently singular events in the same areas with the same three thousand year disparity!

Cave paintings found in the Accacus Mountains in the Sahara Desert in Libya dating from 10,000 years ago show a pleasant, populated, fertile land of rivers and forests that was teeming with game. This is evidence of a pluvial period in North Africa that did not end until 2,500 BC.

Between Libya and the Sudan the archaeologists Di Caporiacioa and Almasy discovered carvings of giraffes, ostriches, buffalo, bulls, zebu-like animals and men with bows and arrows. These were around 10,000 years old. These are in the Arkenu Massif in Libya.

There are carvings of elephants, rhinoceroses, ostriches and crocodiles in Fezzan in Libya that were part of the local fauna before the area became desert ten thousand years ago.

In 1959 chance led to the discovery in a cave in Sonora Province in Mexico of thirty well preserved mummies dating back ten thousand years and belonging to an unknown civilization. These creatures were alleged to be six foot six inches to eight foot tall. There were other reports of giant skeletons being found in caves here in the nineteen thirties. Another report states that in a cave in Sonora ten mummies were found of a race that was totally unlike any hominoid being ever encountered on this planet yet all the same and belonging to the same unknown race.

Around 8,000 BC the continental shelf around New Zealand was dry land with forests, rivers and mountain ranges. Suddenly this continental shelf which was previously forest land rapidly became submerged.

The Baba Yaga Crater in North Korea is ten thousand years old approximately and is sixteen kilometres across. The crater is submerged west of Nampho.

The West coast of Peru rose from the sea to a level of 2,000 metres around 10,000 years ago.

The Cordilleras of the Andes rose to 19,000 feet about 10,000 years ago.

The Andes Mountain range was further raised to its present height leaving the original coastline as a string of salt lakes that still contain sea salt. These are, from north to south, Lake Titicaca, Uyuni, Coipasa and Chiguana in Bolivia, Atacama, Punta Negra and Pedernales in Chile and Arizaro, Piponaco and Hombre Muerto in Argentina. How else did a string of salt lakes get to the top of a mountain range? To make it worse the salt coastal line is now crooked relative to the surface of the lakes. The area was raised and then tilted.

The seafloor fifty miles west of Collao in the Pacific Ocean plunged to a depth of 9,000 feet and submarine exploration has discovered ruined columns and walls at this depth in the Milne-Edwards Deep. Has the sea-floor dropped here after a civilization had built upon it in the remote past?

The sea receded away from China to a huge extent and massive tidal waves hit Ancon in Peru as well as the rest of the western South American coastline. This was the tsunami of the millennia? Or one of many in this period?

Very unusual remains were found on top of the Andes at Acostambo in Hunacavelica Province in Peru. Giant fossilized oysters have been found here nearly two miles above sea level in the Andes Mountains of Peru. These were bivalve ocean dwelling mollusks, *plagiostoma giganteum*, and reached a width of twelve feet with a weight of 650 pounds. They were found by Cuban paleontologist Arturo Vildoza in January of 2001. The estimated age of the fossils is 200,000,000 years but how sure are we of these dates? There were over five hundred of them! The fossils were spread over a wide area and were found closed indicating that they had not been eaten or had died a natural death as oysters open at death. The fact that they were closed indicates that they were prevented from opening by burial in silt and earth. Were these oysters actually only buried during the last cataclysm when the mountains rose?

Beachline fishing communities at Paracas on the west coast of Peru were suddenly raised up four hundred feet.

The Philippines, Sumatra, Borneo and Java were together connected to the landmass of continental Asia. This was the now submerged region of Sundaland.

The larger meteoric impact crater near Frombork in Poland is ten thousand years old and two hundred and fifty metres in diameter. There is a second crater here as well.

There was also another meteorite impact area in Poland from this same area as well. The Morasko Meteorite Reserve is on the northern edge of the Polish city of Poznan. There are seven meteor craters here and the reserve covers an area of fifty-five hectares. Metorite fragments have been found here weighing up to three hundred kilogrammes. The largest crater is one hundred metres across and is around eleven metres deep. Five of the craters contain lakes. Other sources state that the impacts here occurred five thousand years ago. Morasko and Frombork are one hundred and seventy-six miles apart.

There are submerged river canyons on Continental Shelfs from the Rhone River, the Loire and the Seine in France, the Tagus in Portugal as well as from now nonexistent rivers in North Africa that head out into the Atlantic Ocean. A bulletin of the Geological Society of America commented that these sunken river canyons suggested that world wide lowering and rising of the sea levels amounting to 8,000 feet must have occurred since the late Tertiary age, namely the Pleiocene Era, the age of man. The land must have dropped as the sea levels have never been that high.

With sea levels reduced to their Ice Age levels there would be extensive lands reaching out thirty miles or more into the Atlantic from Portugal, Spain and Morocco. This would give in excess of 8,000 square miles of habitable coastal plains in the Gulf of Cadiz and northern Morocco alone.

Read all of this before? You probably have. I did not date anything here. I only collected them. And what you are about to read next seems to be an improbable collection.

What can you see here? Not that much as the points are too small to recognize. To the right are the huge Chelkar-Aralskaya Crater and the much smaller Togyz Crater. The middle cluster of yellow pointers are the Russian impact group from the same period. To the north is the Vanelahti Crater. The three isolated points on the left are the two Polish impacts and the one French one that we know of. In the bottom middle is the El Fayum Depression in Egypt. What the Hell, literally, was happening in this period? Let the games in Russia begin!

The Lezhninskoe Lake is actually a meteorite crater that was created ten thousand years ago near Kirov in the Pizhanka region of Russia. The lake is perfectly round and has high lakesides that are ten metres high. The lake is thirty-eight point six metres deep.

The Vanelahti Crater in Karelia in Russia is ten thousand years old and is a phenomenal 82.5 kilometres in diameter.

And you are still saying that nothing happened in this period? Russia was being bombarded in this period!

Which period though? Was 8,000 BC and 9,000 BC and even 10,000 BC being used as generic general dating because no one had any specific date? How could we have several major impact events in the same area over such a limited time period? Do we have to condense the time period?

Chukhlomskoye Lake was formed by an impact crater ten thousand years ago. The lake is four kilometres from Nozhkon in the Kostromskaya Oblast in Russia. The crater is nine kilometres in diameter. Ancient local legends mention a flaming princess and her chariot crashing into the lake. There were obviously human witnesses to the event and it stayed in their memories for ten thousand year!

The Kostromskiye Razlivy near Kostroma in Russia is an impact crater 36.4 kilometres in diameter. The crater is ten thousand years old.

The Krugloe Lake in Krasnoyarsk Krai in Russia is an impact crater created ten thousand years ago. It has a diameter of one kilometre. Ozero in Russian means lake.

The Zabore Ozero in Leningradskaya Oblast in Russia is ten thousand years old and 2.2 kilometres in diameter.

All of these lakes in this text are classified as impact craters caused by asteroids, meteorites or comets or cometary fragments.

Beloe-Bordukovskoe Ozero in the Moscow Oblast in Russia is 660 metres long by six hundred metres wide. The crater is ten thousand years old.

Chernoye-Bordukovskoe Ozero in the Moscow Oblast in Russia is 670 metres long by five hundred and thirty metres wide. The crater is ten thousand years old.

The Eryomkovo Lake near Moscow is a ten thousand year old meteor crater that is one kilometre wide.

Lake Karpovskoe near Shatura in the Moscow Region is an impact crater formed ten thousand years ago. The lake has a diameter of two hundred and seventy metres and is 31.4 metres deep. Melted impact glass has been found here as well as moissanite crystals which are formed in conditions of high pressure.

Lake Lemeshenskoe in the Shatura Region of Moscow Oblast near Toshal is an impact crater lake that is three hundred and fifty metres long by two hundred and ninety metres wide. The crater is ten thousand years old and there is a ringed hill here.

The Muromskoe Ozero in Moscow Oblast in Russia is much larger than the other impact created lakes nearby. The lake is 5,250 metres long by 2,840 metres wide.

Nerskoye Ozero is a lake in a meteorite crater in the Moscow Region. It is ten thousand years old and 1.29 kilometres in diameter. It is near Ozeretskoye.

The Orekhovo-Zuevskiye Crater in the Moscow Region is 1.6 kilometres in diameter and is ten thousand years old.

The Sherninskiye Ozero Crater in the Moscow Region is nine hundred and fifty metres in diameter and is ten thousand years old.

The Smerdyachee Lake or Smelly Lake impact crater is elliptical and has a well developed rim wall fifteen to twenty metres high unlike other lakes in the same area. The size of the lake is two hundred and fifteen metres long by two hundred and five metres wide. The depth is thirty to forty metres. The lake is one hundred and forty kilometres east of Moscow. Other elliptical lakes in the same region might be impact craters as well. Other sources say twenty thousand years.

Svyatoye Ozero in the Moscow Oblast in Russia is 670 metres long by five hundred and thirty metres wide. The crater is ten thousand years old.

Vvedenskiye Ozero in the Moscow Oblast in Russia is 840 metres in diameter. The crater is ten thousand years old.

Yarmoly Ozero in the Moscow Oblast in Russia is 660 metres long by six hundred metres wide. The crater is ten thousand years old.

The Zolotaya Veshka Ozero Crater in the Moscow Region is 1.24 kilometres in diameter and is ten thousand years old.

Hang on haven't we met some of these lakes before? Yes, around three thousand years before. What is the coincidental value of two massive meteorite showers hitting the same small area in Russia in three thousand years? Nil would probably be the best answer. What would be the statistical probability of it being a single meteorite shower resulting in a mass of impacts at the same time? This would be a certainty. And the mixing up of the dates? That is now expected as well. We also have impact reports from one thousand years before as well. This is too much for statistical probability.

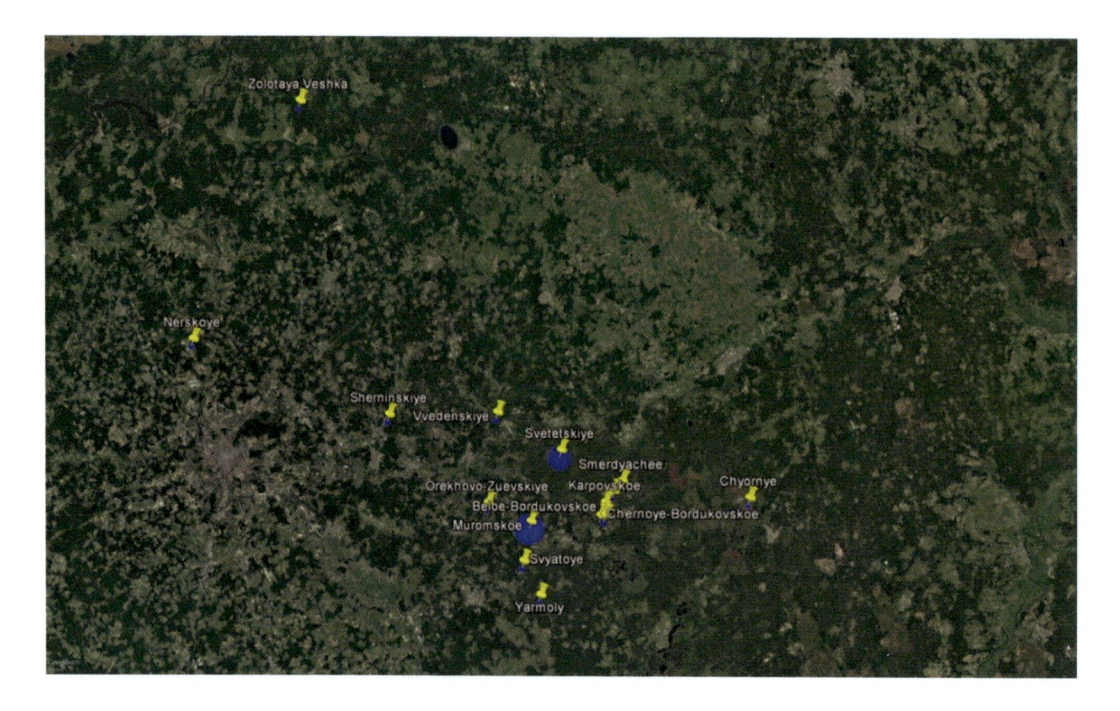

How many impact craters do you want? These are the ten thousand year old impact craters near Moscow in Russia. What did the sky look like in this period as they came crashing down? Were they parts of a single rubble asteroid that broke up as it hit our atmosphere? Rubble asteroids? A rubble asteroid is not a single monolith but a mass of numerous pieces of rock that have coalesced under the influence of a gravitational field. These rubble piles or rubble asteroids form when an asteroid or moon is smashed to pieces by an impact. The shattered pieces fall back together due to self-gravitation. These asteroids have low densities and are prone to explode in the atmosphere whilst descending. Many asteroids and even several planetary moons are now believed to be orbiting rubble piles. Even comets are regarded as forms of rubble and ice held together by their own gravitational fields.

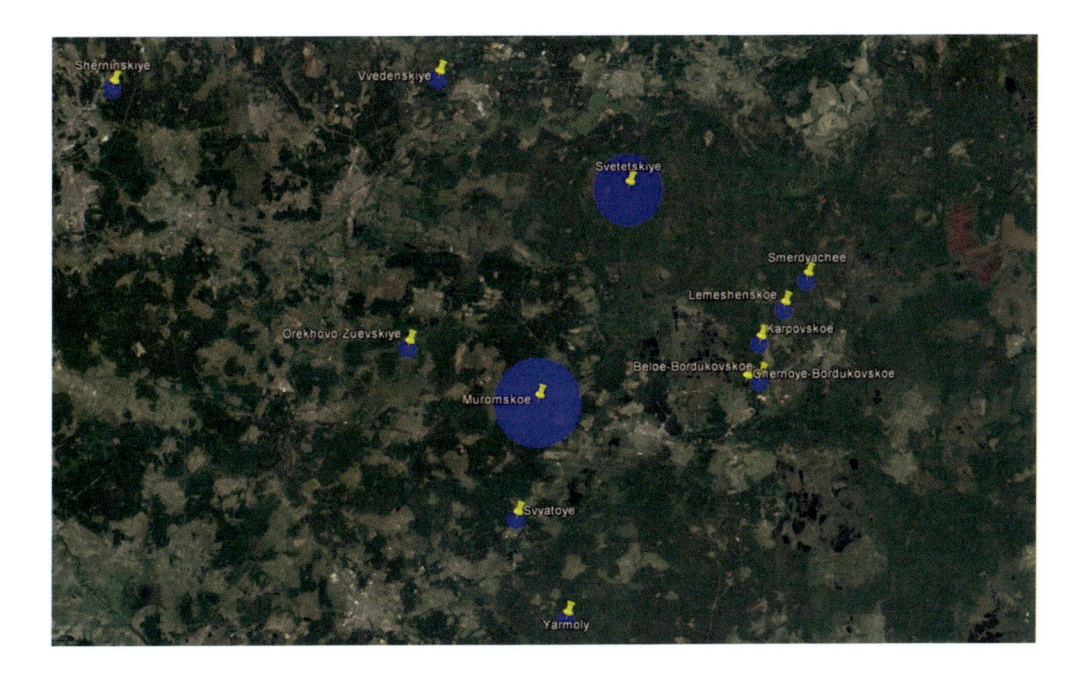

We head further towards the right-hand portion of the Moscow region crater field.

The Svetloyar Ozero in East Nizhny Novgorod is an impact crater two hundred and ten metres long and one hundred and seventy-five metres wide. It was created thirteen thousand years ago. Svetloyar Ozero is in East Nizhny Novgorod in the Gorkiy region of Russia.

The Borovoye Ozero near Moscow in Russia is an impact crater four hundred and seventy metres across. It was created thirteen thousand years ago.

Chernogolvka 1 and Chernogolovka 2 are meteoric impact craters in the Moscow Region in Russia. Number one is three hundred metres wide and number 2 is fifty metres wide. They were formed thirteen thousand years ago.

Lake Lukovoye is formed by a thirteen thousand year old impact crater that is six hundred and forty metres wide in the Moscow area in Russia. Around 10,900 BC there were massive impacts in Russia. Was this impact part of this meteorite or asteroid shower? One hundred years is nothing in geology.

The Orlevo Ozero impact crater is around thirteen thousand years old and is three hundred metres in diameter. Orlevo Ozero is near Orlikovo northeast of Moscow in Russia.

The Svetloe Ozero near Moscow is an impact crater three hundred metres across. It was created thirteen thousand years ago.

What sort of meteoric storm was occurring in this period?

Russian impact craters in the right-hand portion of the Moscow region crater field.

The Ozerskoe crater is eleven thousand years old and 480 metres long by 200 metres wide. It is in Nizhegorodskaya Oblast in Russia.

The Izyar Lake Impact Crater also in Nizhegorodskaya Oblast in Russia is ten thousand years old. The crater is 370 metres long by 180 metres wide.

The Kuzmayar Lake Impact Crater in Nizhegorodskaya Oblast in Russia is ten thousand years old. The crater is five hundred and forty metres in diameter.

The Nestiar Ozero or Lake in Nizhegorodskaya Oblast is an impact crater that is ten thousand years old and six hundred and twenty metres in diameter.

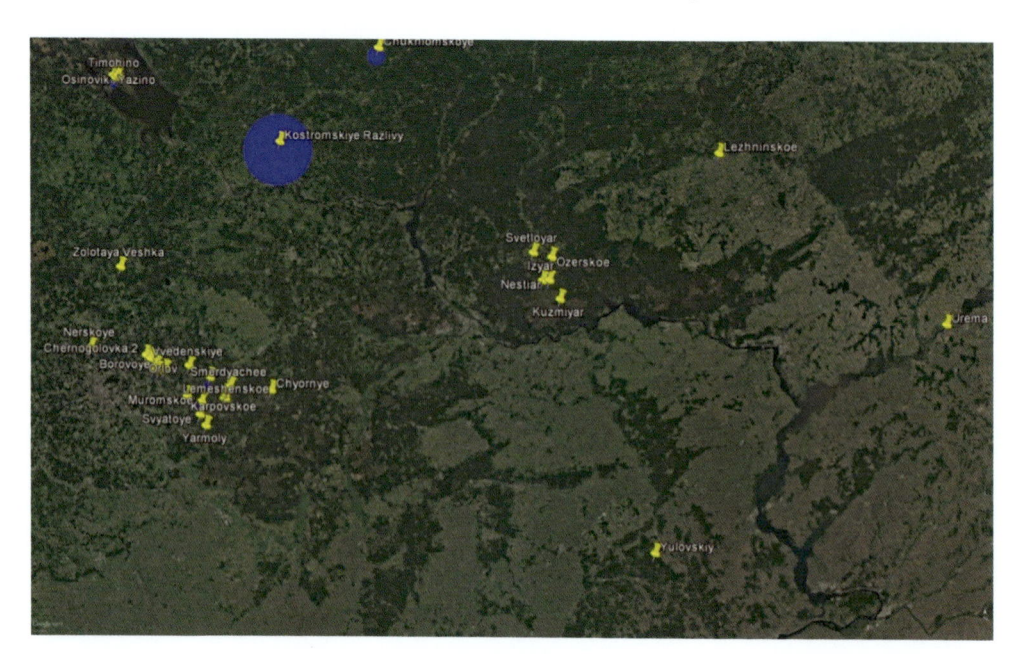

Impacts in Nizhegorodskaya Oblast in Russia to the east of Moscow in this same period.

Now we move to another area. The Tigoda Ozero in Novgorodskaya Oblast in Russia is an impact crater that is ten thousand years old and eleven kilometres in diameter.

Valday Crater or Lake in Novgorodskaya Oblast in Russia is an impact crater that is ten thousand years old and eleven kilometres in diameter.

Vershinskoye in Novgorodskaya Oblast in Russia is an impact crater that is ten thousand years old and 10.6 kilometres in diameter.

The Virovno Ozero in the Novgorod Region is ten thousand years old and 6.9 kilometres across.

The Zamoshskoy Ozero in the Novgorod Region is ten thousand years old and 8.2 kilometres across.

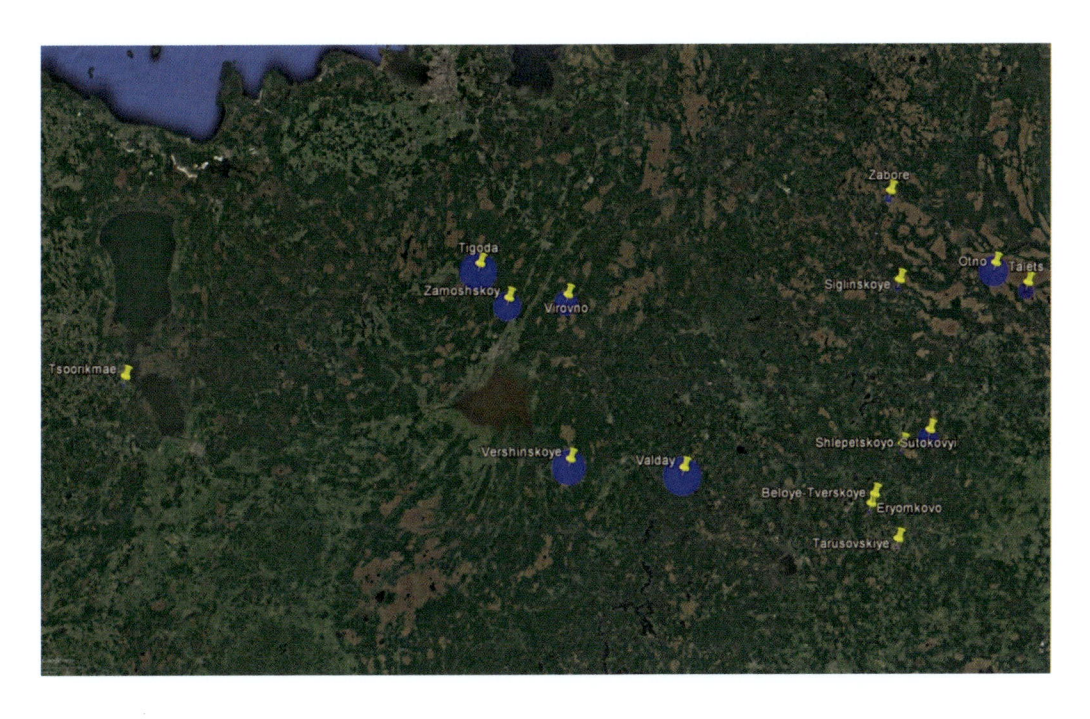

Impacts in Novgorodskaya Oblast in Russia around 8,000 BC.

There are numerous circular crater lakes in this area that are around ten thousand years old. Beloye-Tverskoye is 1.2 kilometres in diameter and is in the Tverskoye Oblast.

The Shlepetskoyo Ozero Crater in the Tverskaya Oblast of Russia is ten thousand years old and around seven hundred metres in diameter. Shlepetskoyo Ozero is near Tyukhtovo in Tverskaya Oblast in Russia.

The Sutokovyi Crater in the Tverskaya Oblast of Russia is ten thousand years old and 6.5 kilometres in diameter.

The Tarusovskiye Ozero in the Tverskaya Oblast of Russia is ten thousand years old and 11.3 kilometres in diameter.

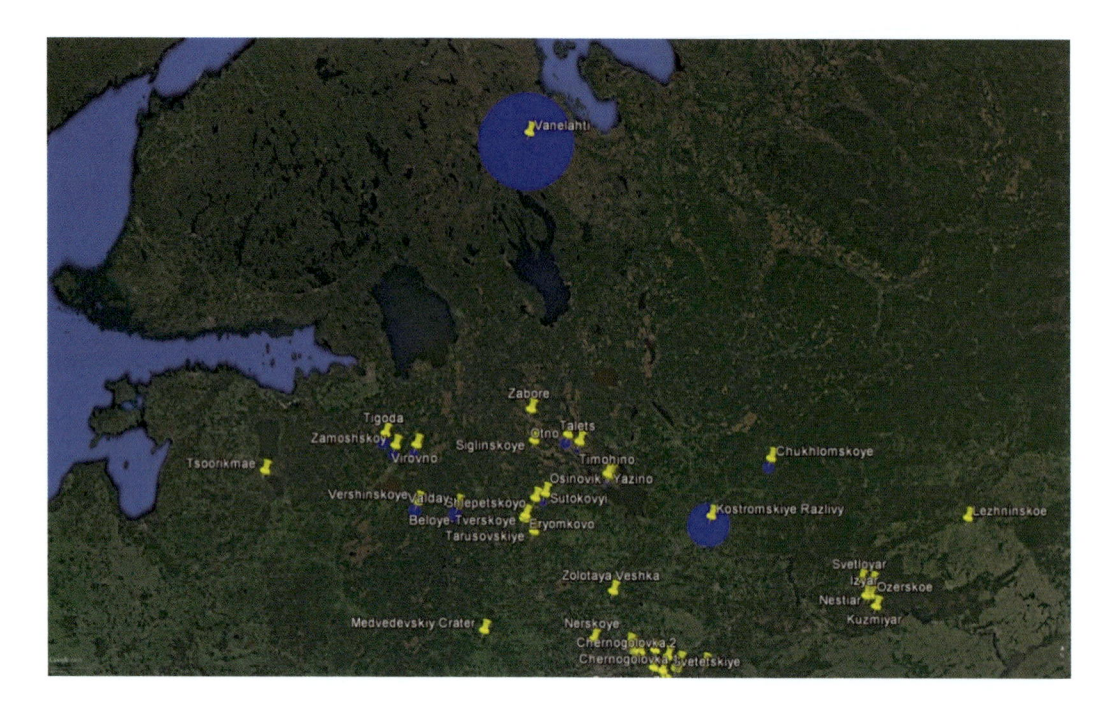

Impact craters in Tverskaya Oblast, Vladimirskaya Oblast and Vologdaya Oblast in Russia.

The Chyornye Crater in Vladimirsk in Russia is six hundred and sixty metres in diameter and was created ten thousand years ago.

The Svetetskiye Crater in Vladimirskaya Oblast in Russia is four kilometres in diameter and was created ten thousand years ago.

Urema Lake near Nizhnekamsk in Tatarstan in Russia is ten thousand years old and 1.1 kilometres in diameter.

The Srednikovo Crater in the Ulyanovskaya Oblast in Russia is ten thousand years old and 1.7 kilometres in diameter.

The Yulovskiy Ozero in the Ulyanovskaya Oblast in Russia is ten thousand years old and 870 metres in diameter.

The Otno Ozero in the Vologodskaya Oblast in Russia is ten thousand years old and 9.1 kilometres in diameter.

Yazhino, Timokhino and Osinovik Craters in Vologdaya Oblast. How does normal science explain this clustering of craters? Especially during the largest meteorite storm to ever hit the Earth! No wonder the ancients commemorated this with their obsession with astronomy.

The Siglinitsy Ozero in Vologodskaya Oblast in Russia is ten thousand years old and 2.6 kilometres in diameter.

The Talets Ozero in Vologdaskaya Oblast in Russia is ten thousand years old and 5.5 kilometres in diameter. This Ozero or Lake is surrounded by other obviously crater-formed Ozeros as well.

Are the circular and elliptical lakes in this photo also impact craters? They even overlap the same as the Carolina Bays in North America.

Allowing for geological variance can I ask if the Russian impact craters that are dated to 11,000 BC and 10,900 BC might even be from this same period as dating is very confused in this period. This would make the impactoid showers in this period the most spectacular ever seen by the human race.

This map gives you an indication of how widespread the meteorite shower or storm was. From Siberia to France. And these are only the known ones. How many more are under the seas or lost in deserts or mountain ranges? Or just not even noticed. Just another lake in a land of thousands of lakes.

Do we really need words here now? What sort of impact stream was this? Other than an enormous one!

Which way was the impact stream heading?

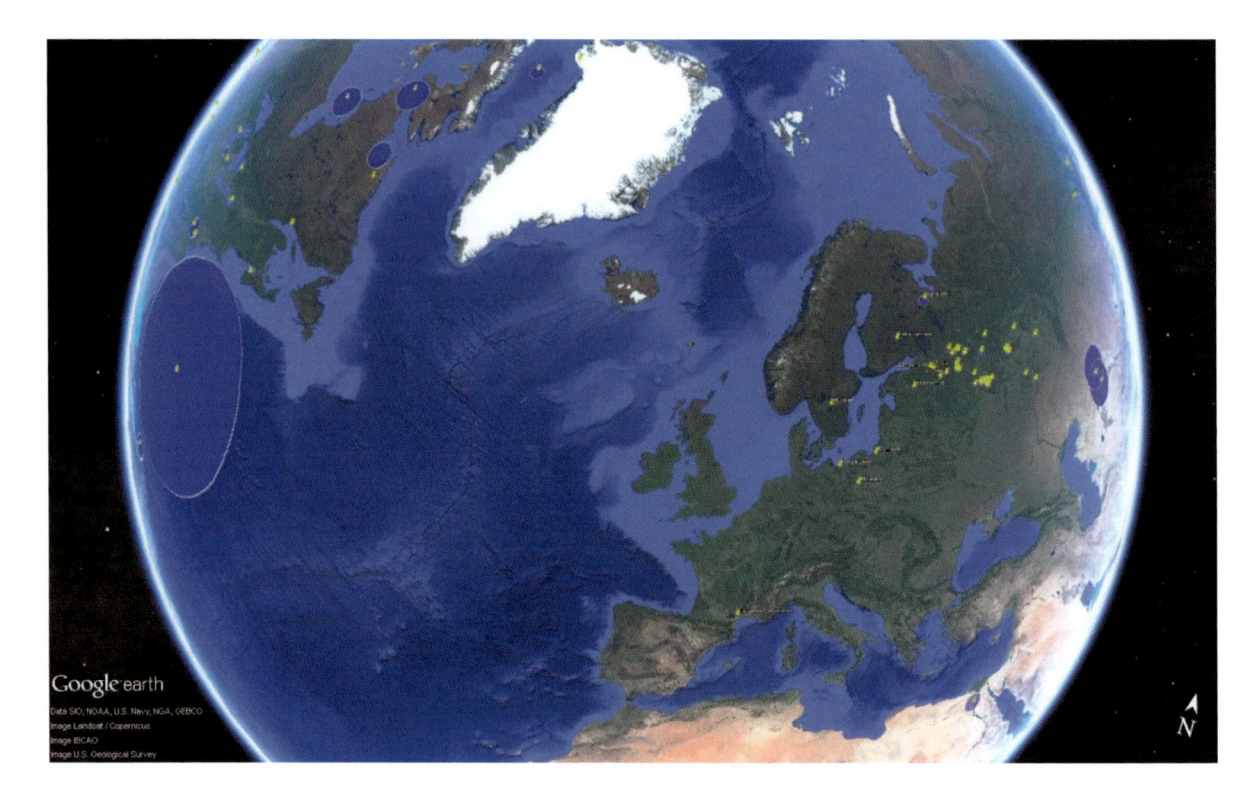

In the above graphic we have the Bermuda Impact to our left and impact craters to the right and impact craters in North America to the north east. Did the impactoid that hit Bermuda split into two parts that headed into two different directions as theyfragmented? There is a lot of ocean in between in which one could hide many undiscovered craters or even one other large one on the Mid-Atlantic Ridge which we have discussed earlier. Or had several massive impactoids hit the earth in the same period? How could two impact streams two thousand years apart hit the same areas with pinoint accuracy? They had to be the same cataclysmic swarm of meteors.

The Sahara Desert is one of the most inhospitable environments on Earth. Yet tiny crocodiles live in isolated desert pools that they could not have reached by migration. Ancient skeletons of elephants, giraffes, antelopes and other animals have been found, thousands of miles from their present homes in the fertile parts of Africa as well as human remains, flint tools, rock paintings and engravings that are thousands of years old. The rock paintings depict herds of cattle, hunting, racing and dancing. At some time in its history the Sahara was fertile land. In exploring dry canyons cut thousands of years ago by deep rivers Scientists have found Old Stone Age tools dating back 500,000 years. The new Stone Age began in the Sahara between 8,000 BC and 6,000 BC when the climate became rainy. At this time the Sahara was very fertile. There were very organized communities there. Around 2,000 BC the Sahara began to dry up again.

The Sahara, a relatively young desert was green savannah until the tenth Millennium BC. This savannah was brightened by lakes boiling with game and extended across much of Upper Egypt. The delta area further north was marshy but dotted with many large and fertile islands. The climate was significantly cooler, cloudier and rainier than today. For around 2,000 to 3,000 years around 10,500 BC it was almost continuous rainfall. Then the floods came. After the floods the arid conditions set in. This period of desiccation lasted until 7,000 BC when the Neolithic subpluvial began with one thousand years of heavy rains followed by 3,000 years of moderate rainfall which proved ideal for agriculture. For a time the deserts bloomed and supported communities that they cannot do today.

In November 1981, a radar scan was taken from the "Columbia" space shuttle over the Sudan and Southwest Egypt at an altitude of 125 miles and covering a width of over thirty miles. Computer enhancement showed the beds of buried rivers some as large as the Nile and all flowing south and west, possibly to a large inland sea. All these rivers had tributaries and streams to them suggesting that thousands of years ago the area supported forests and grazing lands for animals. The last time that this area of the Sahara which includes part of Libya, Chad, the Sudan, Egypt and most probably Tunisia and Algeria enjoyed enough water and rainfall to support animal and human populations was about 10,000 years ago.

Under the Sahara desert is the vast underground Sea of Albienne covering over 230,000 square miles. Scott Elliot states that the Sahara was not formerly part of the Mediterranean but a huge inland lake, which some Earth convulsion in recent times turned into a desert. Early writers describe it as a wooded area with great rivers supporting a dense population and various fauna including antelopes, giraffes, elephants, lions and panthers. The mountains and plateaus rose as great islands.

The change in this appearance was brought about by sharp variations of temperature and later by the actions of wind and water. Pastoral lands encroached more and more onto what had been jungle and the process of drying was accelerated up by the grazing of livestock.

The depression formed by the Shotts, shallow marshy lakes of Algeria and Tunisia were below sea level and would be flooded if a series of protected coastal dunes were ever removed.

In 1926 Claude Roux argued that in post-Glacial times most of North Africa was under the sea and the mountains of Morocco and Algeria constituted a peninsula. Eventually the land rose or the sea fell and the seas and lagoons dried up leaving the present day deserts and salt marshes.

In the post-glacial Quaternary period also according to Claude Roux North Africa was a fertile peninsula that was bounded by great shallow lagoons stretching from the Atlantic and the Mediterranean to the southern Atlas Mountains. This mountainous stretch of land was very fertile and thickly populated with animal and plant life abounding. Eventually the lagoons receded towards the coast leaving lakes and salt marshes and the onslaught of the desertification. The lakes and salt marshes are the Schotts and Sebkas.

A stratigraphic layer of fused green glass has been found here that could only have been caused by a nuclear explosion or massive heat sources from above. Dates for this green glass and the aerial explosion that caused it range from thirty million years to ten thousand years, it all depends on which dating system that you are using. Peculiar how this green glass only appears in desert areas? Was the origin of the green glass part of the explanation of the forming of the deserts? Did this sand vitrification start the desertification that would become the Sahara Desert?

Does this next discovery throw all that you have ever imagined about mammoths to be totally untrue? This was on the Berezovka River in Perm in Siberia. In 1901 a sensation was caused by the discovery of a mammoth carcass, as this animal seemed to have died of cold in midsummer. The contents of its stomach were well preserved and included buttercups and flowering wild beans. This meant that they must have been swallowed about the end of July or the beginning of August. The creature had died so suddenly that it still held in its jaws a mouthful of grasses and flowers. At this instant it seemed that it had then been caught up in tremendous forces and hurled several miles from its pasture ground. The pelvis and one leg were fractured. The huge animal had been knocked to its knees and then frozen to death at what is normally the hottest time of the year. What throws mammoths around and then snap freezes them in midsummer? Massive tsunamis or horizontal mud slides during a snap-freezing earth tilt!

The Meteor Strike Crater near Star of the Sea Mission in KwaZulu-Natal in South Africa is ten thousand years old and 170 metres in diameter.

A series of pictogram carvings on flat, oval stones have been unearthed at Jerf el Ahmar in Syria. These markings consist of lines, arrows and animals and are among the oldest forms of writing in the World dating back 10,000 years to 8,000 BC. They are an intermediary form between Paleolithic cave art and modern forms of writing.

Located on the west bank of the Upper Euphrates in northern Syria is where the earliest examples of lightly fired clay vessels have been found. These date to around 8,000 BC. This is the accepted view. There is Jomon pottery from Japan that also dates to 7,500 BC. Other Japanese pottery dates back to 16,000 BC though there is even older creation of ceramic objects like statuettes etc well before this in Europe. When were ceramics first invented?

In this period around ten thousand years ago sudden climatic changes seriously effected lake levels across Tibet. Long term aridity then started across the Tibetan Plateau.

The North American West Coast rose from the sea to a level of 2,000 metres around 10,000 years ago.

The only exit for the waters of Lake Michigan was the Mississippi River until 8,000 BC. This was followed by a temporary coalescence of Lake Michigan's waters when they formed a super-lake called Lake Algonquin with its only seaward flow being a now vanished channel called the Ottawa Outlet which emptied into the St Lawrence River.

There was a huge glacial lake in what is now western Utah and parts of Nevada called Lake Bonneville that covered an area of 50,000 square miles up until ten thousand years ago. This lake is now the Great Salt Lake in Utah and is six times saltier than seawater. The salts in it are sea salts, not freshwater salts, and it is composed of 55 per cent chlorine and sodium 31 per cent by weight of all dissolved matter. Had a massive tsunami from the Pacific washed over the area and deposited the salt the same as in the salt lakes in South America stretching in a long chain south from Lake Titicaca?

Along the coast of Newfoundland and New England there are numerous stumps of trees in the water that indicate that massive forested areas became submerged at the end of the Ice Age.

The skeletons of a species of Tertiary whale called *zeuglodon* have been found in Alabama in many places in deposits that occur after the last Ice Age. There never was ice cover here in the last Ice Age. Did the land fall and then rise or did massive tidal waves lead to the whale's demise?

Dr Dale Guthrie of the Institute of Arctic Biology made the interesting point about the sheer variety of animals that flourished in Alaska before the eleventh Millennium BC. There was an exotic mix of saber-toothed cats, camels, horses, rhinoceroses, asses, deer with gigantic antlers, lions, ferrets and saiga, a hugely diversity of species so different from that of today. Was the rest of the environment also so different? It looks as if countless animals and plants lived in a more temperate climate in Alaska which was hit by a great catastrophe suddenly around ten thousand years ago. Everything was suddenly frozen in mid-motion. For instance when a bulldozer dug into mammoth bodies the smell of rotting meat spread for miles. This was the first time that the carcasses were exposed to the air since they died ten thousand years ago. The Alaskan muck is dark grey sand. Embedded inside it are the frozen remains of twisted parts of animals and trees intermingled with lenses of ice and layers of peat and mosses. Bison, horses, wolves, bears, lions, whole herds of animals were apparently killed together overcome by a common power. Such piles of bodies of animals or men simply do not occur by an ordinary natural means. As well if they were dwellers of the frozen tundra then they would not have been able to descend into it as it would be frozen solid. The ground under these unfortunate animals was not frozen as these animals died in midsummer and were covered in mud.

At various levels stone artifacts have been found frozen in situ at great depths and in association with Ice Age fauna confirming that men were contemporaries of extinct Ice Age animals. Throughout the Alaskan muck there is evidence of atmospheric disturbances of unparalleled violence. Mammoth and bison were torn and twisted as though by a cosmic hand. In one place we can find the foreleg and shoulder of a mammoth with portions of the flesh and toenails and hair still intact and clinging to the blackened bones. Close by is the neck and skull of a bison with the vertebra clinging together with tendons and ligaments and the chitinous covering of the horns still intact. There is no mark of knife or cutting implement if human hunters were involved. The very heavy animals were torn apart and scattered over the landscape, even though some of them weighed several tons. Mixed with the piles of bones are torn and twisted trees piled in tangled groups. The whole is covered by a fine sifting muck and then frozen solid.

Frozen inside the muck of mangled bones and destroyed trees from this period in Alaska there were also found stone artifacts indicating that man was also involved in the mass destruction. Worked flints called Yuma Points were repeatedly found at great depths under the muck. These are similar to those of the Athabascan People who used to live in the Tanana Valley until relatively recently.

All of western Alaska was unglaciated until the end of the last Ice Age. Remember that the continental shelf of Alaska was also above sea level until this period.

A Smithsonian Institute report published in 1958 as a result of American, Soviet and Indian archaeological research indicated that ten thousand years ago the Eskimos lived in Central Asia.

North of Fairbanks, Alaska, around North Star in Alberta, Canada, and in the Yukon Valley deep frozen woolly mammoth remains have been taken from deep in the ground during the extraction of gold with high-pressure pumps and excavators. The deep frozen stomachs contain leaves and grass, which the animals had eaten. The young lay next to the old, the babies beside their mothers. Such quantities of animals cannot have died all at once in a natural way. The animals had died almost instantaneously and were deep frozen on the spot otherwise they would have shown minimal signs of decomposition. In addition 1,766 jawbones and 4,838 metatarsal bones belonging to a single species of bison were found near Fairbanks.

Between eleven thousand and nine thousand BC there were numerous upheavals in the northern regions of Siberia and Alaska around the edge of the Arctic Circle. Uncountable numbers of large animals have been found; many carcasses still intact as well as astonishing quantities of perfectly preserved mammoth tusks. Hundreds of thousands of individual creatures must have frozen immediately after death otherwise the meat and ivory would have spoiled. The mammoth meat appears so fresh that it has been offered in restaurants in Fairbanks, Alaska, and has been used to feed sled Dogs.

In the mid 1940s Dr Frank C. Hibben, Professor of Archeology at the University of New Mexico went to Alaska to look for human remains. Instead he found miles of muck filled with the remains of mammoth, mastodon, bison, horses, wolves, bears and lions. As bulldozers pushed the half melted muck into sluice boxes for the extraction of gold many animal tusks and bones rolled up in front of the blades like shavings before a giant plane. The carcasses were in all attitudes of death and most of them were pulled apart by some unexplainable disturbance as quoted by Hibben in 1946. Hibben also said that the evidence of the violence of nature combined with the stench of rotting carcasses was staggering and the ice fields containing these remains stretched for hundreds of miles in every direction. Trees and animals were mixed together with layers of peat and moss, mangled together like some giant mixer had jumbled them some 10,000 years ago and then froze them into a solid mass. Trees aren't common on the tundra either. In fact they don't exist. These animals were in forested land. Where is the snowcovered tundra?

There was a cataclysmic tidal wave near Cripple Creek in Alaska approximately 10,000 years ago that left a mass of jumbled bones, ivory and wood. Huge herds of mammoths, rhinoceroses and other animals that were grazing the then temperate plains of Alaska were killed by the tidal wave.

Where the Koyukuk River Valley joins the Yukon in Alaska the mangled remains of a multitude of animals are all smashed together along with the fragments of crushed trees. The frozen bones are of mamoths, mastodons, super-bison and horses and appear as if they were all swept along by a devastating tidal wave. The same happened in the Tanana River as well as the Kuskokwim River that flows into the Bering Sea.

The Kuskokwim River flows into the Bering Sea in Alaska. The mangled remains of a multitude of animals are all smashed together here along with the fragments of crushed trees. The frozen bones are of mammoths, mastodons, super-bison and horses and appear as if they were all swept along by a devastating tidal wave that came from the ocean. The same happened in the Koyokuk River Valley as well as the Tanana River Valley.

Where the Tanana River Valley joins the Yukon in Alaska the mangled remains of a multitude of animals are all smashed together along with the fragments of crushed trees. The frozen bones are of mamoths, mastodons, super-bison and horses and appear as if they were all swept along by a devastating tidal wave. The same happened in the Koyokuk river Valley north of the Yukon River and which actually flows into the Yukon as well as the Kuskokwim River that flows into the Bering Sea.

The remains of Vero Man were found at Vero Beach in Indian River County in Florida in 1915. At least five individuals were found as well as numerous other fossils of animals. Stone artifacts as well as an incised proboscidean tusk, a mammoth tusk, were found. Some reports state that pieces of broken pottery were found as well. This was allegedly too early for pottery but not the only occurrence of pottery found in this period. Projectile points, awls and pins were found here as well as at Melbourne nearby from the same period where the remains of humans were also found.

In 2009 a carving of a mammoth or mastodon was found on a piece of bone found just north of Vero Beach. The bone is heavily mineralized and anatomically correct.

A massive bone yard has been discovered near Gainesville in Alachua County in Florida that contains the largest concentration of fossils ever found in Florida. It is ten thousand years old. Remains of a small, extinct, pronghorn antelope have been found here and normally it is not found east of Central Texas. Other sources state that this event occurred two million years ago. Both times are the accepted dates for extinction events. But which extinction event? And why is there the great disparity in dates?

There is a submarine field of dead sea-elephants off the coast of Georgia dating from the period of 10,000 years ago.

In Georgia marine deposits containing walrus skeletons have been found at altitudes up to 240 feet.

Sixteen kilometres southwest of Marquette in Marquette County in Michigan spruce and tamarack trees were found in growth position between six and eleven metres below the surface. There were distorted strata including sheared trees and intercalcated till. The trees were radio-carbon dated as ten thousand years old. The wood was remarkably well preserved with only the bark and less than an inch of the outer layer showing any carbonization.

Lake Lahontan at its peak around 12,700 years ago had a surface area of 8,500 square miles centred on the Carson Sink and included Pyramid Lake where it was 900 feet deep and five hundred feet deep over the Black Rock Desert. By 8,000 BC it had largely disappeared as had numerous other megalakes around the world. Bones of horses, elephants and camels have been found in the Lahontan sediments. A spear-point was also found in these sediments as well.

At Astor Pass a large gravel pit from the period of Lake Lahontan was found to contain the skeletal remains of *felix atrox* a species of lion extinct since the last Ice Age. The remains of the same species of lion were also found in the La Brea tar pits making them contemporary to each other.

Some reports state that there was definite proof of a large community in Jericho in the West Bank by 8,000 BC. By then massive stone walls thirteen feet thick and ten feet high were encircling an area of ten acres. Stone city walls and towers were built later at Jericho in 7,000 BC. There was an encircling stonewall with a stone tower thirty feet high. Inside the stone tower was an internal spiral staircase thirty feet of which was still standing ten thousand years later. Shallow underground dwellings were common. This is the same as at Maadi opposite Giza in Egypt. The first tower in the Mediterranean area was built at Jericho and is thirty feet high. There are also walls sixteen feet high that were replaced by walls 22 feet tall. Skilled engineers must have done this work yet they did not know pottery at all and ate off plates and dishes made of flint and used stone vessels where we use crockery. Their knives, scrapers, saws and augurs were made of flint or obsidian. The houses were shaped like halved eggs and were generally two storeys, the walls being of oval bricks and a floor of burnt stucco. The corners were rounded to avoid collecting dust.

In 8,000 BC the inhabitants of Jericho were constructing enormous fortification walls, gouging out vast trenches in the hard bedrock and erecting a gigantic stone tower in defense against an unknown enemy. Engineering projects such as this require a great amount of social structure and coordination as well as social stability.

Ten human skulls were found here. The features of the dead were modeled with traces of colour and with shells for eyes. These skulls were buried under the floors of the houses and had been made by modeling plaster over the skulls to produce lifelike effects.

The oldest remains from Jericho show that Natufians occupied the area around 10,000 BC because it was a natural spring and oasis during the drought period of the Younger Dryas. Soon there was a large community here covering four hectares when the average for a community was less than one hectare. This was comprised of beehive houses densely packed together and separated by courtyards and narrow alleyways. This village huddled behind a massive stone wall with a masonry tower bordered by a rock-cut ditch nearly three metres deep and over three metres across.

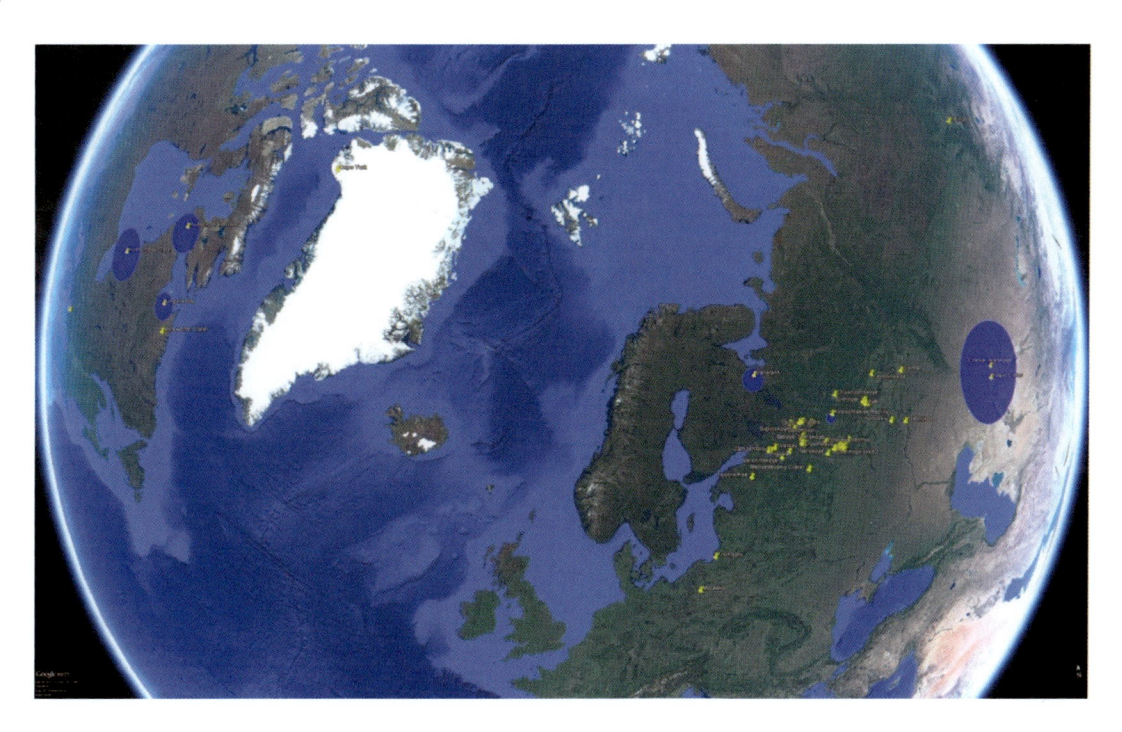

You explain what was happening here! Irrespective of whether we were heading east to west or west to east something enormous was happening. And modern education conveniently ignores it.

In the period of 7,750 BC the water level of the Black Sea was one hundred metres lower than it is now. Incidentally so were the sea levels in this period due to the absorption of water by the glacial ice sheets.

In 1996 the remains of a man 9,700 years old were discovered on Prince of Wales Island in Alaska. The ancient skeleton was radio carbon dated to this age.

Around 7,640 BC world water temperatures suddenly started rising by 4.5 degrees centigrade until 3,000 BC when it reverted. The melting ice caused by this rise in sea temperatures caused a sea level rise of ninety to one hundred and twenty metres.

Around 7,640 BC there was a massive peak in nitric acid in the earth's atmosphere. This occurs when vast amounts of nitrogen are burnt up in the atmosphere during a cometary impact. Vast amounts of hydrochloric acid and sulphuric acid are also formed from seawater when a comet hits it. This was deduced from ice cores taken at Camp Century in Greenland in 1980. Which impact event this was is still unknown as dating as you now know can be quite erratic and variable.

There was a sudden increase in radiation around the Earth in 7,600 BC. This could have been caused by massive destruction of the ozone layer by cometary impacts.

In 1970 tektites were found embedded in fossilized wood dating back 9,500 years ago plus or minus 200 years. Can we allow for a little more latitude in our dating as the average dating of ten thousand years ago is a little too loose to be taken as gospel.

Cometary fragments landed in the Tasman Sea ten thousand years ago.

According to some climatologists Egypt and the Sahara Desert were still forested and wet in this period with elephants and giraffes roaming the area as well as aquatic animals in large lakes. This was not the Egypt of today. Around 2,500 BC the Great Pyramid would still be surrounded by lawns and water filled canals.

Britain and Ireland were were still connected to Europe. Suddenly England was separated from Europe in this period by a suddenly emerging North Sea.

Yes, I know, it sounds like I am repeating things but it is all due to the geological dating variance. If we can have variances of millions of years, a couple of thousand years is absolutely nothing. It is probably better to just compress this period into one as geological dating is not known for its finesse.

According to some climatologists Egypt and the Sahara Desert were still forested and wet in this period with elephants and giraffes roaming the area as well as aquatic animals in large lakes.

The Irish Sea also flooded the land between Ireland and Britain. Maybe this is the real date?

Around 7,600 BC the climate in Scotland warmed up to become almost Mediterranean.

Around 7,600 BC there was a Neolithic community existing in what is now Castle Street in Inverness in Scotland. They had built their community on top of a layer of white sea sand that had been found in various places in Scotland including Inverness and Fife. This was residue of an enormous tsunami that had washed over Scotland.

In North Wales geologically recent seashells have been found at heights up to four hundred metres up the slopes of mountains such as Moel Tryfan that could have only been deposited by a massive tsunami. North Wales was covered in seawater. Had the tsunamis raced across the Atlantic Ocean and flooded Wales? Were these the same tsunamis that had scattered sea sand over Scotland?

Edith Kristan-Tollman and Alexander Tollman of the University of Vienna postulated that about 9,500 years ago, 7,600 BC, seven large fragments of a comet crashed into the Earth. Fragments of a piece of this comet crashed into the Otz Valley in Austria. The Tollmans, also called Tollmann, stated that it was actually 3.00 AM on September 23rd 7,553 BC. The Tollmans argue that the large oceanic impacts of cometary fragments caused the legends of Noah's Flood. The Tollmans were this precise in their dating due to a Middle Eastern tradition that the event occurred on the day of the Autumn Equinox. The Tollmans based their research on tektites that were laid down in sediment around 7,600 BC.

The Tollmans stated that the cometary fragments landed almost simultaneously in the Tasman Sea between New Zealand and Australia, in the South China Sea, in the west-central Indian Ocean, in the North Atlantic, in the central Atlantic south of the Azores Islands, in the Pacific Ocean off Central America and again in the Pacific Ocean west of Tierra del Fuego. The story goes that these impacts generated tsunamis, triggered volcanic eruptions and earthquakes and created much damage.

The Tollmans stated that the Kofels Crater in the Otz Valley of the Austrian Tyrol was created by the cometary or asteroid impact.

Incidentally a very strong acid layer in Greenland ice cores was dated to 7,630 BC. Not too far off and the radiocarbon content of tree rings indicated a major event occurring in 7,553 BC, the same year as that postulated by the Tollmans.

Were these part of the same meteor shower that had hit the Tasman Sea near Australia as well as Vietnam?

Most interesting discoveries were found in Valdivia in Los Lagos Province in Ecuador. Research showed that though the Jomon people of Japan had developed pottery around 7,500 BC the Valdivian pottery showed Jomon styles that were contemporary with it. Specific traits shared only by the Valdivian pottery and the Jomon pottery were a unique form of vessel rim and a stone neck rest. There were also similar clay model houses. This pottery is similar to that found in the coastal plains of Georgia and South Carolina in the same period. Had it diffused from Ecuador or been independently created? Did stranded Japanese or Jomon fishermen arrive in Ecuador and introduce different styles of pottery to the Valdivians or were there already established trade links via the Great Circle Route on the Japan Current which skirted north-eastern Siberia and the Aleutian Islands and then went south? Remember that the oceans and seas were the superhighways of the ancients.

Incidentally the San Pedro Culture in Ecuador dating back to seven hundred years older than Valdivia shows no Jomon characteristics.

Later ceramics found on the coast at Valdivia north of Guayaquil in Ecuador have been found that date back to 3,000 BC. Some dated to 3,620 BC and the original discoverers believed that they were related to ceramics found on the coast of Peru at Ancon and Guanape. The Valdivia pottery though resembles early pottery from Kyushu in Japan that had been created by a long lost race called the Jomon of whom little is known. The finishes showed an astonishing degree of similarity in surface finish, decorative techniques, vessel shape, motifs and rim treatment which confounded matters even more. The Jomon pottery though dated back to 7,500 BC and had evolved over the centuries to the level where it was the same as the Valdivia ceramics. No New World antecedents have been found for the Valdivian style of pottery. Incidentally there is a Pacific Ocean current that runs across the North Pacific Ocean from the east coast of Japan to the west coast of Ecuador. Is this a coincidence? Were the early inhabitants of Japan also colonists of Ecuador? The Valdivians cultivated maize around 3,300 BC.

A most amazing discovery was found in the Gulf of Cambay or Gulf of Khambat in Gujarat State in India. The remains of a city under the water have been found 25 miles off Surat on the western coast of India. The area was submerged no later than nine to ten thousand years ago. The remains are at a depth of 120 feet and cover an area of five miles by two miles. Debris recovered from the site including construction material, pottery, large sections of walls, beads, sculpture and human bones and teeth that have been carbon dated to be nearly nine thousand five hundred years old. Inscribed evidence of an unknown language was also found. The ruins are evidence of extensive human habitation along what was a river prior to 7,500 BC. Yes, you have read this before. There are date discrepancies with some of our data. It all depends on which expert dated it.

Lake Lauricocha is near the Bolivian border with Peru. In a cave at Lake Lauricha in Huanaco in Peru the remains of human skeletons have been found that are dolichocephalic, or elongated in skull formation dating to at least 7,500 BC. The Peruvian engineer Augusto Cardich discovered a culture very high up in the Andes in this area near Lake Lauricocha that is thirteen thousand years old, 11,000 BC.

At Shakta Cave in the Pamirs at an altitude of 14,000 feet the Russian Archaeologist V. A. Ranov found drawings made with red mineral paint depicting a bear, boar and ostrich, none of which can survive in the present Arctic temperatures. A clue to the age of the drawings, the highest altitude prehistoric drawings in the World, was found at Markanus where settlers left artifacts and ash. The latter were of burnt cedar and birch, which do not grow in the region today. Carbon-14 dating showed 9,500 years or 7,500 BC. This is regarded as the ninth deepest cave system in the world. The cave is actually the Pantjukhina Cave near Shakta Vjacheslav in Abhkazia in the Pamir Mountains of Siberia.

Cometary fragments landed in the Tasman Sea between Australia and New Zealand and created tektites around 7,500 BC.

At Ashikli Hoyuk near Aksaray in Turkey extremely long beads made of hard stones, seven on the Mohr hardness scale, have been found. These are of agate, carnelian and quartz and were up to 5.5 centimetres in length. We need highly specialized diamond tipped tungsten carbide drills to cut these and the drills have to be constantly cooled by running water. The beads date to between 7,500 BC and 7,000 BC and are the same as those found in Tell Abu Hureyra in Northern Syria. In 1989 a burgundy agate necklace consisting of ten oval and butterfly-wing-shaped beads all between 2.5 and 5.5 centimetres in length were found here by Archaeologist Ufuk Esin.

What is all of this early archeology to do with impact events? For one it provides a welcome relief from a tedious listing of impact craters and it also indicates that there were human witnesses and survivors to these events. Are these impacts and tsunamis the origins of the numerous legends and mythologies from the past?

Tektites found in Vietnam date to 7,500 BC and were caused by an impact in the China Sea. Around 9,500 years ago, 7,500 BC, there was a major oceanic impact possibly of a comet in the Tasman Sea southeast of Australia as well as in the China Sea near Vietnam. Were there two comets or asteroids or meteors? Were these the "Falling Stars" of legend?

In Spirit Cave in Churchill County in Nevada in 1940 the body of a man was discovered. The body stank with a musty, thick sweetness that clings to the back of your throat according to one witness. The body was lying on a fur blanket, dressed in a robe made from twisted strips of rabbit pelt and hemp cords, wearing leather moccasins and with a skillfully woven twine mat sewn around his head and shoulders. Another mat was placed beneath him. There was still skin left on his back and shoulders as well as a tuft of straight, dark hair. Spirit Cave man was about 45 years of age when he died and 1.75 centimetres tall or 5 ft 10 inches. Initially thought to be two thousand years old it was found that he was in fact over seven thousand years old after radio carbon dating much later when better dating techniques were available. The site dated back to 7,400 BC.

Thirteen miles east of Fallon, Nevada, the Wheelers, working for the Nevada State Parks Commission, were surveying possible archaeological sites to prevent their loss due to guano mining. Upon entering Spirit Cave they discovered the remains of two people wrapped in tule matting. One set of remains, buried deeper than the other, had been partially mummified (the head and right shoulder). The Wheelers, with the assistance of local residents, recovered a total of sixty-seven artifacts from the cave. The mummy was approximately 9,400 years old, 7,400 BC, older than any previously known North American mummy. The mummy exhibits Caucasoid characteristics resembling the Ainu of Japan, as well as a possible link to Polynesians and Australians that is stronger than to any Native American culture. Had this person due to his Ainu similarities sailed to North America using the North Pacific Current that flows from Japan to North America?

The unusually complete skeleton of a Caucasian man was found in Kennewick in Benton County in Washington dating to 7,300 BC. The man's head and shoulders were mummified preserving much of the skin. The body was discovered on July 28th 1996, when a skull with teeth was discovered in the Columbia River shallows during the annual hydroplane races. Other bones were found at the site after the authorities arrived and it appeared eventually that the man had been deliberately buried. A CAT scan showed a Clovis Era spear point lodged in his pelvis that had healed over. The remains though were of an individual with a distinctively long and narrow skull, a projecting nose, receding cheekbones, a high chin and a square jaw indicating that he was a modern Indo-European, Polynesian or Japanese Ainu rather than the round face, prominent cheekbones and round skull of the modern American Indians or Siberians. His lower bones were long compared to the upper bones, traits uncommon among modern American Indians. The only bones missing were the sternum, a few ribs and some of the small bones of the wrists and the feet. The remains were dated between 9,500 and 9,300 years ago around 7,300 BC. It was the skeleton of a mature male about five feet nine inches tall and aged about 45 years old. He had been heavily muscled, apparently suffered chronic pain from badly worn teeth and there was the three inch long spear point embedded in his pelvis. The new bone had grown around the Clovis Era spear point indicating that Kennewick Man had lived for quite some time after being speared. Kennewick Man did not look Asian either or of Siberian origins. The skeleton was taller than other ancient human remains found in the Pacific Northwest. His face and skull were narrower and his jaw was not very prominent when compared to modern day Amerindians. Kennewick Man's remains had been postulated to be of a Caucasian but bear more resemblance to those of the Ainu of Japan or the Polynesians of the Pacific who originated in south east Asia.

Some historians including Thor Heyerdahl promoted the theory that the Polynesians arrived in the Pacific via the Japan Current which hits the west coast of North America and then curves west again. It is almost impossible to sail a ship west to east in the South Pacific but the currents of the South Pacific flow east to west almost constantly. This was the route the Spanish Treasure Galleons used to take to get back to Spain from South and Central America via the Philippines due to this difficulty.

In May, 1976, a forgotten civilization was found at Ban Chiang in Northern Thailand near Laos dating back to 7,260 BC. The culture was farming rice and keeping draft animals as early as 5,000 BC and were using metal alloys as early as 3,600 BC and possibly 1,500 years before that or 5,100 BC. More recent research in Ban Chiang indicates that bronze was being created here as early as 5,000 BC. During the sixth to seventh millennium BC Ban Chiang was already a fully developed agricultural community.

Incidentally rice was being grown in Thailand between 7,260 BC and 5,520 BC, well before its development and cultivation in China. Had it been developed in Sundaland?

A visual roundup of 11,000 BC to 7,000 BC.

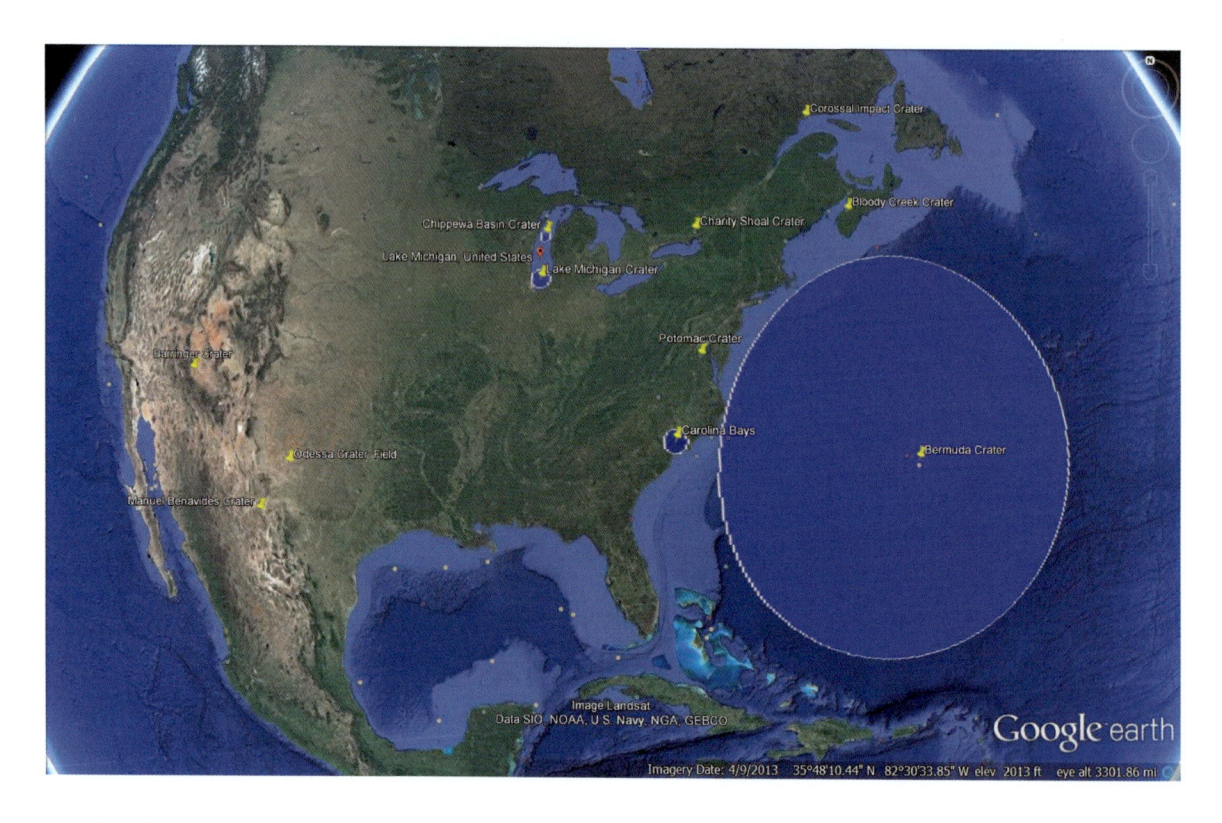

Major impacts of ten thousand B.C. To the right is the Bermuda Crater. Next to it are the Carolina Bays. Unshown is the Puerto Rico Trench which is due south of the Bermuda Crater. In Lake Michigan are the Lake Michigan Crater and the Chippewa Crater. To the left are the Barringer Crater, the Odessa Craters and the Manuel Benavides Crater. To the top right are from north to south, Corossal Impact, Bloody Creek, Charity Shoal and the Potomac Crater.

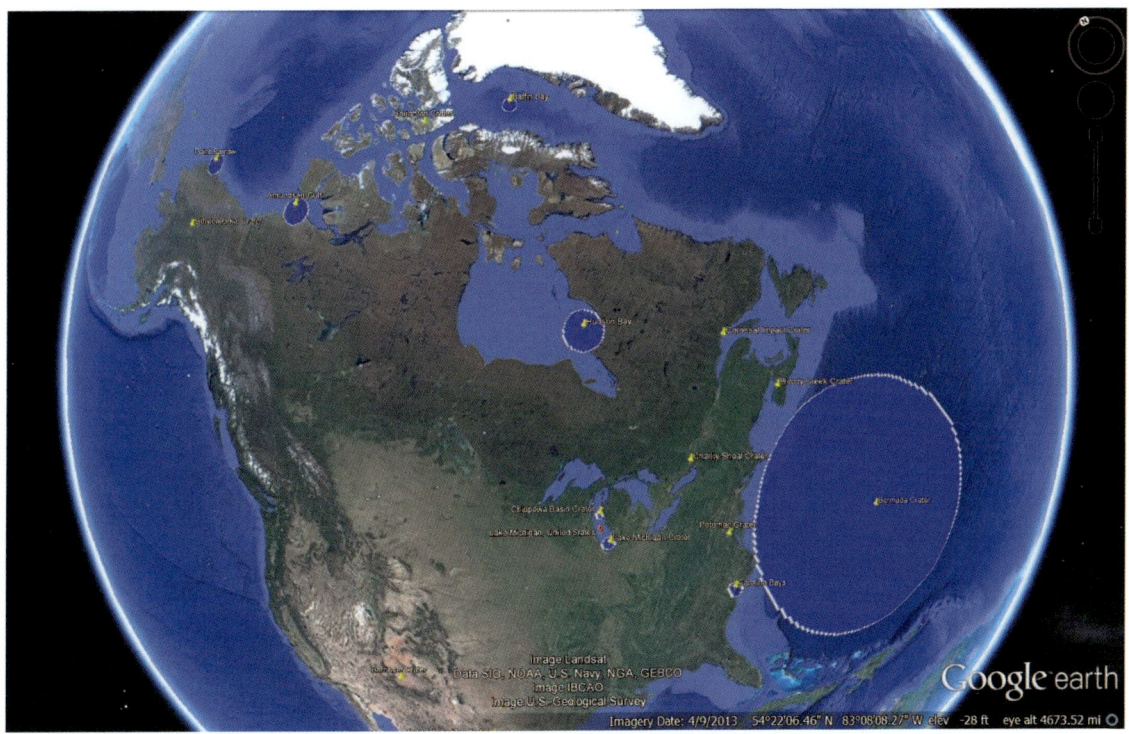

Another view of North America shows the Hudson Bay Crater. To its north and to the northwest are Sithylemenkat Crater in Alaska and Point Barrow. As we go east in what appears to be an arc we come across Amundsen, Houghton and Baffin Bay.

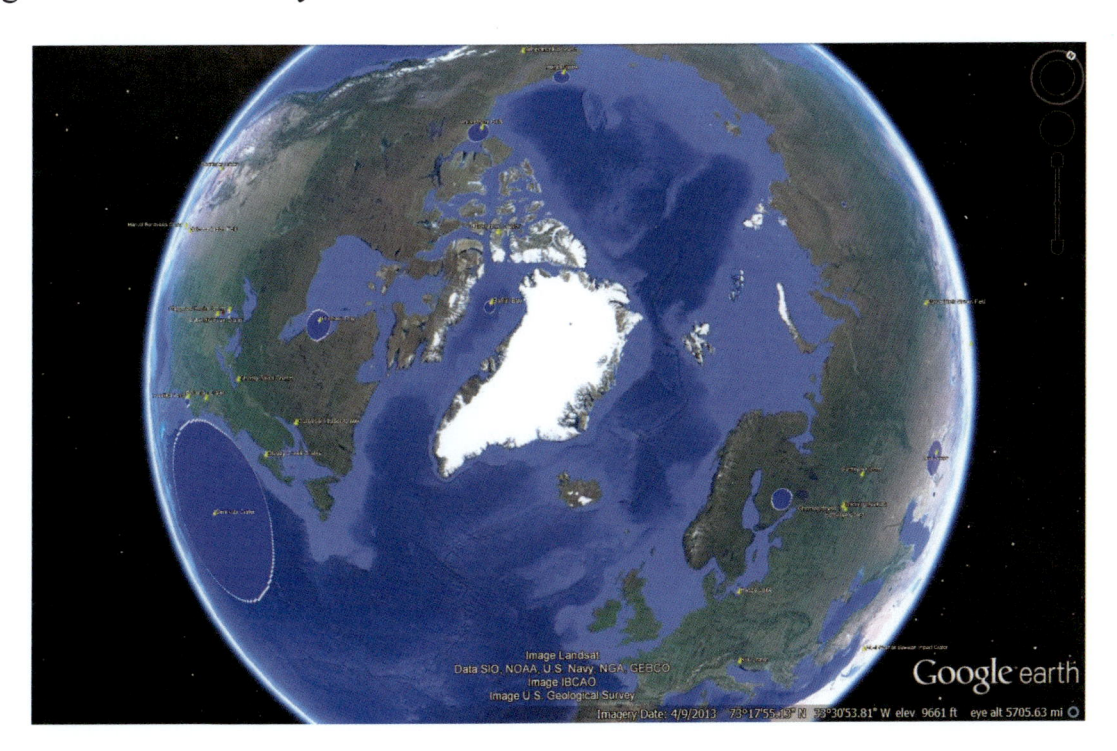

Another scene of North America showing an arc of impacts from Point Barrow, running through the Amundsen Crater, over Hudson Bay and then lining up with the Bermuda Crater. To the right are other impacts from the same period.

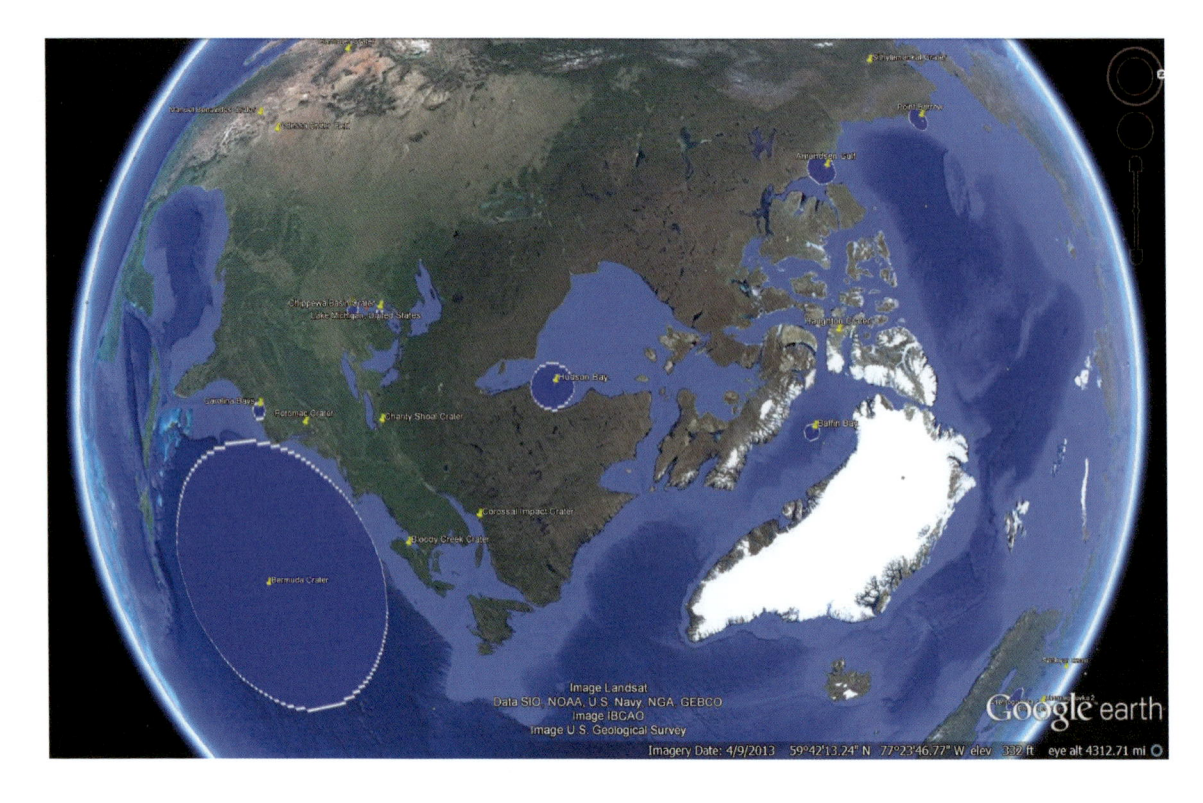

Here is another view of the same North American impacts seeming to form a straight line. Was this a cluster of impacts or one major impact breaking up? What are we to make of this? You never read of it in the history books! This is major but hidden history that effected all of us!

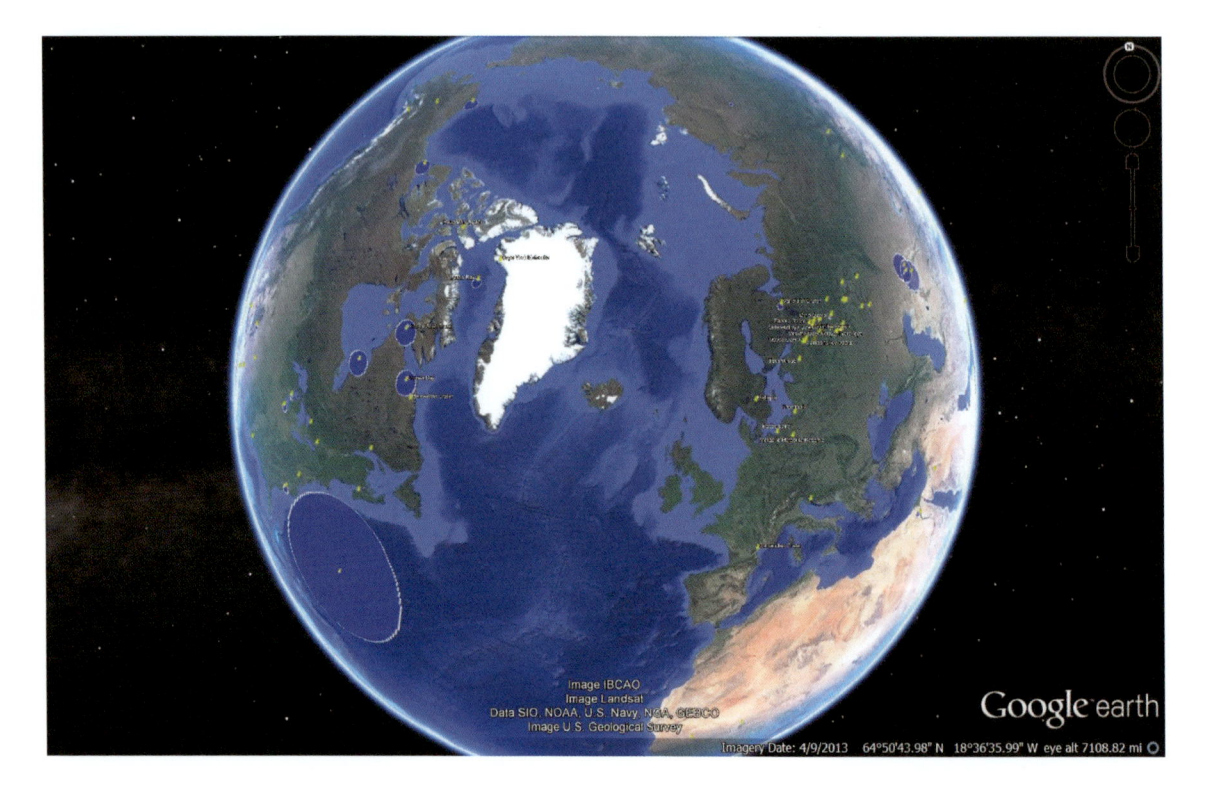

The big blue area to the left is the Bermuda Crater at the bottom of an impact stream across Canada and to Alaska. To the right is the Russian mega-strewn field that appears as if a bucket of gravel was thrown at the Earth covering a very large elliptical area. Allowing for the nature of newly discovered rubble asteroids the cosmic body that struck the right-hand part of the above map seems to have fragmented. Did Earth have an impact with a binary pair of cosmic bodies in this period or was it a loosely bonded together peanut-shaped rubble asteroid?

Scientists were originally puzzled by the densities of some asteroids as they appeared less dense than the meteorites that composed them and that in fact had come from them and were in fact pieces of asteroids.

Large interior voids are possible in these asteroids because of their low gravity. Many of them may even have what appears to be a fine and solid-looking regolith or covering on the outside but the asteroid's gravity is so weak that the friction between fragments dominates and prevents small pieces from falling inwards and filling up the voids.

As an example the asteroid Itokawa appears to be two contact binaries, that is two asteroid bodies touching, with or without rubble filling the boundary between them. This is peanut-shaped the same as some recently observed comets. How well held together are comets then?

Phobos, the largest of the two moons of Mars, is now considered to be a rubble asteroid.

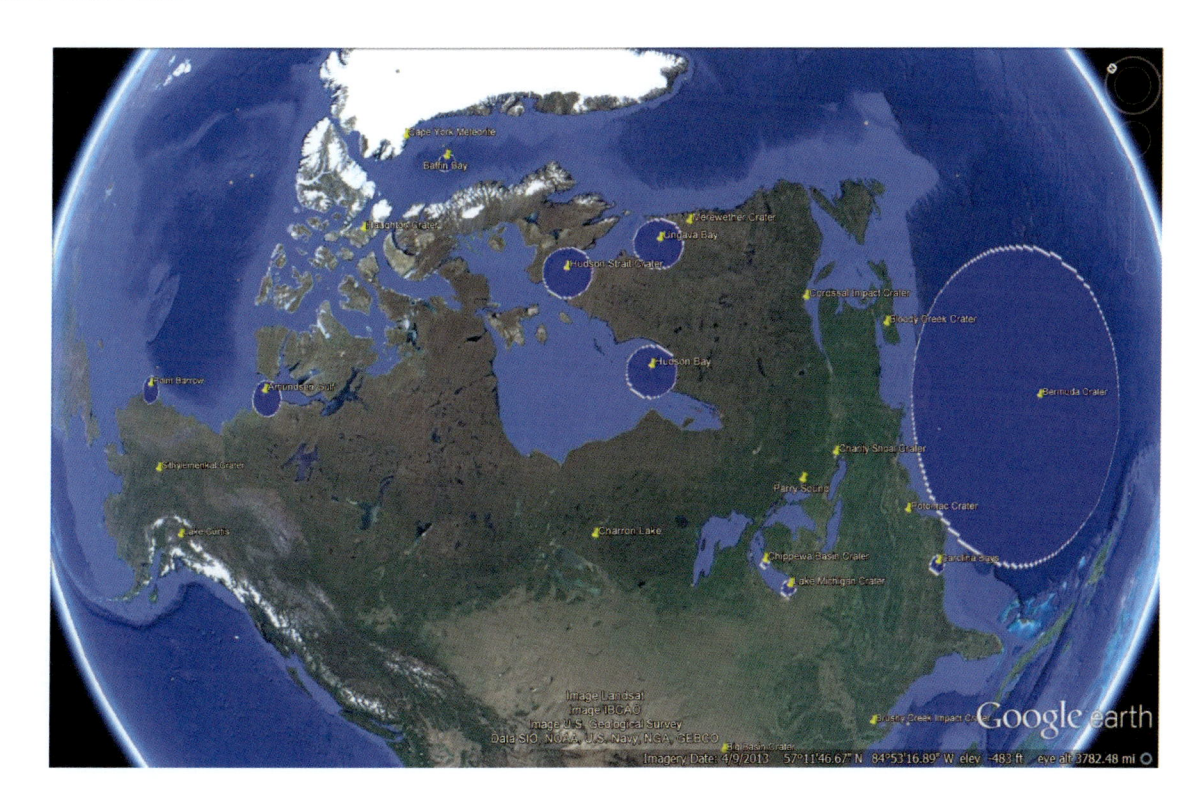

You tell me what happened here then? The blue shaded areas indicate the crater sizes of the impacts? This seems to be a row of them in diminishing sizes.

What is the apparent tally for the Eleventh Millenium?

The tally of meteorite impacts in the Eleventh Millennium. We had the undiscovered Labrador Crater which spread impact detritus over a huge area, the Aral Impact Crater in Kazakhstan, the Svetloyar Ozero in East Nizhny Novgorod, the Borovoye Ozero near Moscow, Chernogolvka 1 and Chernogolovka 2 near Moscow, Lake Lukovoye also nearMoscow and the Orlevo also in the Moscow region, the Potomac Crater in Virginia, the Luanda Crater in Angola, the Marcador Paleolagoons in Bolivia, the Botswana Paleolagoons, the enormous Brazilian paleolagoons, Hudson Bay, the Corossal Impact Crater in Canada, the Charity Shoal Crater in Canada, the Manuel Benavides Crater in Mexico, the Novosibirsk Crater Field in Siberia and the Kaoma Craters in Zambia.

What then is the score for the Tenth Millenium? We have major impacts in the Bahamas, the Bermuda Impact, Beni and Araona in Bolivia, Amundsen Gulf, Baffin Bay, Bloody Creek Crater in Newfoundland, Hudson Bay, Lake Saimaa in southern Finland, Lake Racze on Wolin Island in Poland, meteorite craters found in the Carolinas, Georgia, Florida and Virginia, Sithylemenkat Crater near Bettles in Alaska, Barringer Meteor Crater in Arizona, Chippewa Basin at the deepest part of Lake Michigan, Lake Superior, Nebraska Bays, Carolina Bays, Puerto Rico, Odessa Craters in Texas, Zhongcangxiang Crater in Tibet, Luna Zeel in India, Ouro Ndia in Mali, Gogui Crater in Mauritania, Zhongcangxiang Crater in Tibet, Point Barrow in Alaska, Crestone crater in Colorado, Mullsjon Crater in Sweden. And these are all craters over half a kilometer in diameter.

Was this an unusual year? Or were there three unusual milleniums?

What meteoric events happened in the Ninth Millenium? The tally is the Morro de Cuero Crater in Argentina, the Curtis Lake Crater near Palmer in Alaska, the Brushy Creek Impact Crater in St Helena Parish in Louisiana andthe Parry Sound Impact Crater in Ontario in Canada. Not many really when compared to other milleniums.

Then we have the impacts for the eighth Millenium. For the period during the Eighth Millenium there were numerous impact events. Far more impact events than you would ever expect. The Rio Cuarto Crater Field in Cordoba Province in Argentina, the Joey Crater in the Southern Ocean south of Australia, the Kangaroo Crater in the Southern Ocean south of Australia, the Flinders Impact Crater in Bass Strait, the Hickman crater in Western Australia, tektites found in Victoria produced by an impact southeast of Tasmania, Otz Valley in Austria , the Merewether Crater in Labrador, the Hudson Bay Crater, the El Fayum or El Fayoum Depression in Egypt, the Tsoorikmae Impact Crater in Estonia, the Cabrerolles Crater in France, the Cape York Meteorite in Greenland, the Kukla Crater in the Southern Ocean,the Togyz Crater and the the Chelkar-Aralskaya Crater in Kazakhstan, the Baba Yaga Crater in North Korea, the Frombork and Morasko craters in Poland and Meteor Strike Crater in South Africa. In Russia there was the Lezhninskoe Lake, the Vanelahti Crater, the Chukhlomskoye Lake, the Kostromskiye Razlivy, the Krugloe Lake, the Zabore Ozero, the Beloe-Bordukovskoe Ozero, the Chernoye-Bordukovskoe Ozero, the Eryomkovo Lake, Lake Karpovskoe, Lake Lemeshenskoe, the Muromskoe Ozero, the Nerskoye Ozero, the Orekhovo-Zuevskiye Crater, the Sherninskiye Ozero, the Smerdyachee Lake, the the Svyatoye Ozero, the Vvedenskiye Ozero, the Yarmoly Ozero, the Zolotaya Veshka Ozero, the Svetloyar Ozero, the Borovoye Ozero, the Chernogolvka 1 and Chernogolovka 2 craters, Lake Lukovoye, the Orlevo Ozero, the Svetloe Ozero, the Ozerskoe crater, the Izyar Lake, the Kuzmayar Lake, the Nestiar Ozero, the Tigoda Ozero, the Valday Crater, the Vershinskoye Crater, the Virovno Ozero, the Zamoshskoy Ozero, the Beloye-Tverskoye Ozero, the Shlepetskoyo Ozero, the Sutokovyi Crater, the Tarusovskiye Ozero, the Chyornye Crater, the Svetetskiye Crater, Urema Lake, the Srednikovo Crater, the Yulovskiy Ozero, the Otno Ozero, the Yazhino, Timokhino and Osinovik Craters, the Siglinitsy Ozero and the Talets Ozero.

What is the meteorite tally for the Seventh Millenium? The Lake Hamilton Crater on the Eyre Peninsula of South Australia, the Weepra Park crater near Elliston on the Eyre Peninsula of South Australia, the Big Basin Crater in Clark County in Kansas and the Malha Crater Lake near Kutum in Sudan.

We now have massive evidence of impacts in the four thousand years from the Eleventh Millenium to the Eight Millenium. And then it went quiet again as you see in the Seventh Millenium though it seems obvious that two of the four impacts from this period might have actually fallen in the previous Millenium.

And we have never been told about any of this!

And then we add the Russian impacts to all of this? We do indeed have the sky falling and the seas rising.

7th Millenium BC.

Around 7,000 BC there was a massive electromagnetic surge in the earth's atmosphere that affected the earth's magnetic field. This is an approximate date only with a time error span of one thousand years and could have been caused by a collision with a cosmic body. This event could have occurred in 7,600 BC during the most recent cosmic impacts.

Professor Lhote found rock paintings of prehistoric sheep together with other figures known as Roundheads at Jabbaren in Algeria. These paintings are 9,000 years old around 7,000 BC. The Roundheads have enormous round heads, strange collars, two eyes but no mouth or nose. The members of Lhote's expedition called them the Martians. In the Tuareg language Jabbaren means the Land of the Giants. The height of some of these figures is six metres. They resemble astronauts in space suits. They could also resemble Shamanistic figures and could be open to many more realistic explanations. Professor Lhote found rock paintings in Jabbaren in Algeria of prehistoric sheep together with other figures known as Roundheads. These paintings are 9,000 years old, 7,000 BC. The Roundheads have enormous round heads, strange collars, two eyes but no mouth or nose. The members of Lhote's expedition called them the Martians. In the Tuareg language Jabbaren means the Land of the Giants. The height of some of these figures is six metres. They resemble astronauts in space suits.

Strange drawings were found at Tassili-n-Ajjer in the Sahara Desert in Algeria. The Tassili drawings were discovered by Lieutenant Brenard, or Brenan, in the heart of the Sahara and then examined by Henry Lhote. Besides animal drawings and hunting scenes the sketches show us strange figures, which seem to be wearing space suits with round helmets. Lhote called these figures Martians and said that one of them looked like a man emerging from an egg shaped object covered in concentric circles. There are many thousands of pictures of animals and men including figures in short elegant coats. They carry sticks and indefinable chests on the sticks. Next to the animal paintings is a being in a kind of diver's suit. He is eighteen feet tall. On his heavy powerful shoulders rests a helmet, which is connected to his torso by a kind of joint. There are a number of slits on the helmet where the mouth and nose would normally be. There is the Tassili sphere, which Lhote found on the underside of a semi-circular rock. In a group of floating couples, a woman is pulling a man behind her; a sphere with four concentric circles is clearly visible. On the upper edge of the sphere a hatch is open and from it a thoroughly modern television aerial protrudes. From the right half of the spheres stretch two unmistakable hands with outspread fingers. Five floating figures that accompany the sphere wear tight fitting helmets on their heads, white with red dots and red with white dots. One figure shows a person in overalls tied tightly at the knees and waist and arms and wearing a helmet with only one eye. It resembles a deep-sea diving helmet from the mid-twentieth century. He has no thumbs and as the drawing is of a high artistic calibre the absence of thumbs is a mystery. The helmet is round with two eyeholes and a row of elliptical marks above. The articulation of the neck can be seen. The rest of the garment might be mistaken for coveralls.

The Lake Hamilton Crater on the Eyre Peninsula of South Australia is nine thousand years old and one hundred metres in diameter.

The Weepra Park Crater near Elliston on the Eyre Peninsula of South Australia is nine thousand years old and one hundred and eighty-three metres in diameter. The two craters are fifty-five kilometres apart. What is interesting is that Weepra Park and Lake Hamilton are in line between the Hickman Crater and the Kangaroo Crater, the Joey Crater and Flinders Crater. Did these impacts occur around the same time?

In 1938 Julius Bird, also known as Junius Bird, found a human skeleton in Palli Aike, Palliaiche, on the Straits of Magellan and radiocarbon dating in 1950 revealed that the bones were nine thousand years old, 7,000 BC. The bones were found with the remains of a horse. The skeletons of the men as well as horses were found together in the cave. Muck states that the skeleton is ten thousand years old and was discovered in 1969-1970. Was this a different skeleton? The man of Palli Aike was *Cro-Magnon* according to Muck being tall, muscular, athletic and agile similar to the American Indians of today. Another report states that an elongated human skull was found among bones of sloths and horses. The remains of the man, the horses and the sloths date back to 7,000 BC. The remains were charred.

In 7,000 BC there was a colonization of the Larisa Basin area in Thessaly, northwest of Athens, in Greece. Incidentally this was where legend states that Deucalion and Pyrrha established a kingdom after the Great Deluge. The new inhabitants chose to live on what was left of the old coastal plains that had not been settled by the hunter-gatherers in the country. They chose the floodplains because the soil was light, easily tilled and well watered. There are though no artifacts, no pottery, fabrics or any other type of archaeological remains that would allow an identification of their origins. They appeared to have come by sea and brought their skills with them. Pyrrha was the daughter of Epithemus and Pandora of box fame. What drowned island did they come from then? There would have been quite a few to choose from as the world sealevels suddenly rose on an average of three hundred feet.

Greece during the high Glacial Period was markedly different from the Greece of today. There were huge coastal plains that today are very rare. These lowlands were inundated with the sudden sea level rise.

The Okinawa island chain was part of a narrow continuous peninsula linked to the Chinese mainland until 7,000 BC.

An archeological site at Natsushima in Japan revealed pottery created nine thousand years ago by the mysterious Jomon culture. The pottery pieces were deep conical bowls with cord-mark decoration. This is three thousand five hundred years older than China. Natsushima is near Yokosuka on an island in Tokyo Bay in Kanagawa Province.

The remains of the oldest city at Ugarit, now Ras Shamra, on the Syrian coast are nine thousand years old, dating from at least 7,000 BC. It is situated only 12 kilometres north of the modern port of Latakia. The ruins of Ugarit have been vitrified. This is where the brick or stone has been fused and develops a glasslike glaze. It requires such extreme heat that its origins are unknown. The People of the Sea finally destroyed the city in 1,200 BC during another age of world cataclysms.

On the river Carsamba, thirty miles from the town of Konya and fifty miles from the volcano of Hasan Dag there were twelve cities one on top of each other, the most ancient being nine thousand years old. Catal Huyuk in Anatolia dates from 8,500 to 7,700 years ago, 6,500 BC to 5,700 BC. It was a vast subsurface metropolis discovered by James Mellaart in 1958. Copper objects surrounded by slag-metal were found here dating back to seven thousand BC, 5,000 BC. This indicates that man here had the ability to separate metal from ore and to use fire to shape it. Mortars were found that were used to grind grain. The inhabitants of Catal Huyuk actually cultivated three types of wheat and also planted barley and lentils and grew oleaginous and medicinal plants as well as almond and pistachio trees.

We do not know how they inhabitants of the city polished their incredibly hard obsidian mirrors without leaving scratches or how they drilled fine holes through stone and obsidian beads so thin that a modern needle cannot be pushed through them. There are no predecessors for Catal Huyuk. It appeared as if from nowhere as a fully developed and comfortable civilization. There was an unprecedented command of technology here such as hundreds of knives, daggers, arrowheads and lances of flint or obsidian. Also obsidian mirrors, jewelry and textile work including carpets as well as wooden and basketwork. They did not use pottery. It is as if this culture appeared suddenly in eight thousand BC from nowhere as the other contemporary cultures of the time at Jericho in the Jordan valley or Jarmo in the Kurdish highlands were not this advanced at all. At a level of excavation called VIa the blocks had been fused together by such intense heat that its effects penetrated down to a depth of more than one metre below the level of the floors. This carbonized the Earth, the skeletal remains of the dead, the burial gifts that had been interred with them. The tremendous heat had even halted all bacterial decay. Pieces of carpet have been found in the ruins that are of such a high quality that they compare with the best woven today. These carpet fragments are 8,500 years old, 6,500 BC.

On the north and east walls of shrine V11.14 at Catal Huyuk there is a nine thousand year old painting showing the twin peaks of Hasan Dag with the taller right-hand peak in the midst of an explosion with volcanic rock flying upwards in arching trajectories from the crown. There is also a mushroom cloud of smoke high in the sky above as well as flows of lava rolling down the volcano's ten thousand foot sides. At the foot of the volcano is a village layed out like Asikli on the other side of the Konya Lake. Catal Huyuk was abandoned around 6,200 BC at the bginning of the four hundred year long Mini-Ice Age that ended in 5,800 BC.

Interesting items were found in Coyonu or Cayanu in Turkey. Needles, hooks and scrapers made of cold hammered copper were discovered in southeastern Turkey. They are from 7,000 BC. Two copper pins, a bent copper fishhook and a copper reamer or awl were discovered here indicating advanced metallurgy here in 7,200 BC. Also from Cayonu there was found a piece of cloth wrapped around an antler, possibly to provide a better grip. This cloth is 9,000 years old and is thought to be linen woven from locally grown flax. This is the earliest known piece of cloth in the World. Cayonu was a main bead-producing centre in the early Neolithic period. There are a large number of rectangular stone buildings here with grid plan foundations similar to Nevali Cori. One of these structures had a floor of flagstones into which were set megalithic stones with further standing stones set up in a row nearby. There is another building called the skull building, which is a rounded stone structure 7.9 metres by 7.9 metres with a ruined apse at one end. In two small antechambers some 70 human skulls were unearthed, all of them slightly charred. There were also the remains of some 259 individuals.

Also in Cayonu excavators found a large chamber with an enormous one tonne cut and polished stone block, which acted as an offering table. Nearby a flint knife was found of which microscopic examination revealed residue of human and animal blood.

The people of Dorak in Anatolia in Turkey were domesticating and raising cattle as well as growing cereals nine thousand years ago. They were also making tools, weapons and mirrors out of obsidian, a black volcanic glass.

The Big Basin Crater in Clark County in Kansas is around nine thousand years old and 1.3 kilometres wide.

An amazing discovery was found in Lamos Cave in East Nevada. Professor Luther S. Cressmann of the University of Oregon came across two hundred pairs of sandals woven from fibres in the Lamos Caves. They were over nine thousand years old and were of very high and modern craftsmanship.

At Qualat Jarmo, east of Kirkuk in Iraq millstones for grinding wheat and ovens for baking bread were found here in 1948 by Robert J. Braidwood. They date to 6,750 BC. Robert Braidwood also found various copper items as well as copper oval shaped beads. A single bead of smelted lead was found as well, one of the oldest examples of metallurgy anywhere in the World. Houses with stone foundations were found as well as use of agriculture, domestic animals and stone vessels. There was a city here that is one of the earliest in the world.

At Ishango in the Democratic Republic of Congo which was formerly called Zaire a small prehistoric engraving tool was found near Lake Edward. It was bone handled with a sharp chip of quartz fixed to one end. It was dated to 6,500 BC. The bone handle had a series of scratched markings down its length. The scratches according to Alexander Marshack indicate a record of lunar phases, of the sets of the new, quarter and full moons over a few months.

Around 8,500 years ago the inhabitants of St Helens Island near the town of Las Vegas in Guayas were cultivating a number of crops including maize. The history of the domestication of the maize plant is still a mystery.

The Isle of Man was still connected to Great Britain. The Bristol Channel was dry land as far as what is now Westward Ho.

Cardigan Bays west coast followed a straight line north-south from what is now Anglesay to the Pembroke Peninsula in Wales.

The North Sea achieved its present configuration due to subsidence around 6,500 BC. Prior to this one could walk from what is now Flamborough Head in Great Britain to the mouth of the Elbe River in Germany.

The Malha Crater Lake near Kutum in Sudan was created in 6,290 BC and is nine hundred metres in diameter. The Malha Crater Lake is an impact crater.

6,200 BC. The four hundred year mini-Ice Age.

Greenland ice core samples showed that in the period of 6,200 BC to 5,800 BC there was a ninety per cent drop in methane levels worldwide. This indicated that marsh gas, as a byproduct of wetlands and the production of methane, had dropped considerably in this incredibly dry period. Methane levels are raised when the climate is warm and moist and lowered when it is cold and dry.

Temperatures in the Northern Hemisphere dropped and rains were meager as a wave of aridity swept across southeast Europe, the Ukraine and southern Russia.

During the mini-Ice Age Lake Victoria in Africa responded to Polar cooling and drying by abruptly dropping its water level.

Around 6,200 BC huge meltwater accumulations caused by the warming weather patterns undermined the retreating Laurentide ice sheet in northern Canada and at some time around then the huge Laurentide ice sheet imploded on itself sending a massive outflow of water that cascaded like a landborne tsunami southwards into the Gulf of Mexico. At the same time another freshwater tsunami rushed into the North Atlantic Ocean. This was as strong as the flood caused by the draining of Lake Agassiz five thousand years before.

The Gulf Stream stopped again for up to four centuries and once again much dryer, colder conditions hit Europe again. The moist westerly air masses stopped to be replaced by cold northerly airflows.

The Balkans and eastern Mediterranean area suffered severe droughts. The Laurentide ice collapse caused a rapid rise in the worlds oceans.

Four hundred years of drought hit the Balkan area of the eastern Mediterranean.

During the four hundred year drought and mini-Ice Age many refugees from Anatolia and other areas may have settled by the western and southern shores of the Euxene Lake. During the mini Ice Age temperatures here were considerably warmer and sheltered river valleys still offered fertile well-watered soils. Pollen samples indicate that grasslands and steppes covered the the coastal plains around the lake and the Euxene Lake was like a large fertile oasis. Within six hundred years this area was swamped by the newly created Black Sea when waters from the rising Mediterranean crashed through the Dardenelles barrier into the Sea of Marmara and then through the Bosporus and flooded the area! Thus was created what was to become the Black Sea. This was later in 5,600 BC.

During the mini-Ice Age the area around the Euxene Lake would have been a sanctuary for humans and game as the surrounding mountains of the

Fertile Crescent, the Negev Highlands and the Anatolian Plateau chilled and could not support the populations that had previously lived there. This was due to its setting below the surround sea levels causing it to remain warm. The Euxene Lake held vast quantities of water when the other lakes in the area shriveled up due to the extreme dryness and became salt pans and salt marshes which were unpotable to humans and animals. Streams and rivers from the Balkans, the Alps and the Caucasus mountains kept the waters fresh and in flow year round whilst the Euphrates River had dried up unable to reach Abu Hureyra which was abandoned in this period. Pollen from the bottom of the Black Sea came from cereals and pulses and not trees like oaks and alders. It was good farming country and seeing that the older settlements in the area had been abandoned then it was only logical that the area now covered by the Black Sea was once covered with townships and farms with all of the activities and productivity and organization that this entailed. The previous settlements had already showed stable well planned communities based on agriculture and trade and in this veritable Garden of Eden during the incredibly dry mini-Ice Age they would have prospered just from being in the right place at the right time, until the wrong time came along.

The living witnesses to the Black Sea Flood would have been townspeople, some of them tilling fields, planting seeds and harvesting crops and breeding animals whilst there would also be artisans and merchants. There would have been bricklayers and builders, carpenters, painters and plasterers, sculptors, basket weavers, leather workers, jewelers, potters, bakers and shopkeepers as well as priestly and organizational classes. The average lifespan was thirty years with rare exceptions to sixty years and they suffered from malaria and arthritis. These would have been very organized communities, after all Asikli had four hundred houses alone and Catal Huyuk was sizeable in itself. There could have been even larger communities here in what was to become the Black Sea.

Even today there is a rushing undercurrent from the Mediterranean Sea to the Black Sea in the Bosporus that then travels along a submerged river valley under the Black Sea itself. There is very little sedimentation here unlike the Mediterranean Sea that it adjoins.

The Carioco deep-sea core which was taken from the Caribbean shows evidence of the mini-Ice Age lasting four hundred years.

The English Channel, according to some geologists, was now in full flow between England and France.

Cores drilled into an ancient coral reef in Indonesia show that there was a mini-Ice Age in this period as well as abrupt sea surface cooling by about three degrees centigrade.

Around 6,200 BC the North Sea started rising suddenly by 46 millimetres a year and huge tracts of southern Scandinavia vanished under the rapidly rising sea. Remember that these dates are approximate though they seem to be pretty much on the money so far in this period.

At the beginning of the mini Ice Age the Tigris and Euphrates Rivers were emptying into the Persian Gulf which had recently formed now being only twenty metres lower than modern sealevels. As the surge occurred with the melting of the Laurentide ice masses the waterlevel in the Persian Gulf eventually peaked at two metres above present levels between 4,000 BC and 3,000 BC.

A four hundred year drought started in Anatolia around 6,200 BC. Lake levels fell quickly and many lakes vanished as well as rivers and streams. The oak and pistachio forests retreated across the area as temperatures sank rapidly. Many farming communities were abandoned as sedentary survival techniques became ineffective. Communities once again had to travel to find food to survive.

Deep sea cores taken from the Western Pacific Warm Pool, which has the highest mean sea surface temperatures in the world, show that there was a massive cooling event starting in 6,200 BC that lasted for four hundred years.

What is the meteorite tally for the Seventh Millenium? The Lake Hamilton Crater on the Eyre Peninsula of South Australia, the Weepra Park crater near Elliston on the Eyre Peninsula of South Australia, the Big Basin Crater in Clark County in Kansas and the Malha Crater Lake near Kutum in Sudan.

What happened at the End of the Last Ice Age.

We have looked at sea levels rising and falling.

We have seen the destruction of the ice masses and tidal waves crashing over the earth.

We have seen death and destruction on an unimaginable level.

Twice!

We have seen the total change of the surface of the planet that we call our home.

We have seen instantaneous cataclysms!

There was nothing slow about the events at the end of the last Ice Age.

It was like the Earth was being slapped down by the stars!

From the Forty-third Millenium to the Seventh Millenium there were few major impacts with the Earth. Everything appeared normal. But we had freak periods.

Around the Eighteenth Millenium there was a brief upsurge in impact activity. We have Veevers Impact Crater in Australia, the Iturralde Crater in

Bolivia, Lake Lasnamae in Estonia, Lake Tremorgio in Switzerland and Elco in Nevada. Is this unusual? And these are only the found ones. This is five large impacts in one thousand years. Is this still quite a lot? And don't forget the Australites and tektites showering upon Australasia as well as Southeast Asia, Indonesia and the Philippines.

Then it was quiet again for six thousand years. Then in the Eleventh Millenium it all started again.

The tally of meteorite impacts in the Eleventh Millennium. We had the undiscovered Labrador Crater which spread impact detritus over a huge area, the Aral Impact Crater in Kazakhstan, the Svetloyar Ozero in East Nizhny Novgorod, the Borovoye Ozero near Moscow, Chernogolvka 1 and Chernogolvka 2 near Moscow, Lake Lukovoye also nearMoscow and the Orlevo also in the Moscow region, the Potomac Crater in Virginia, the Luanda Crater in Angola, the Marcador Paleolagoons in Bolivia, the Botswana Paleolagoons, the enormous Brazilian paleolagoons, Hudson Bay, the Corossal Impact Crater in Canada, the Charity Shoal Crater in Canada, the Manuel Benavides Crater in Mexico, the Novosibirsk Crater Field in Siberia and the Kaoma Craters in Zambia.

Why the sudden explosions of meteorite impacts in the Eleventh Millenium? Or do we need to merge the four milleniums from the eleventh to the eighth as the falls appear to repeat at the same places again? Is this too much for statistical reality?

What then is the score for the Tenth Millenium? We have major impacts in the Bahamas, the Bermuda Impact, Beni and Araona in Bolivia, Amundsen Gulf, Baffin Bay, Bloody Creek Crater in Newfoundland, Hudson Bay, Lake Saimaa in southern Finland, Lake Racze on Wolin Island in Poland, meteorite craters found in the Carolinas, Georgia, Florida and Virginia, Sithylemenkat Crater near Bettles in Alaska, Barringer Meteor Crater in Arizona, Chippewa Basin at the deepest part of Lake Michigan, Lake Superior, Nebraska Bays, Carolina Bays, Puerto Rico, Odessa Craters in Texas, Zhongcangxiang Crater in Tibet, Luna Zeel in India, Ouro Ndia in Mali, Gogui Crater in Mauritania, Zhongcangxiang Crater in Tibet, Point Barrow in Alaska, Crestone crater in Colorado, Mullsjon Crater in Sweden. And these are all craters over half a kilometer in diameter.

Then we had a break in the Nine Millenium as it went relatively quiet again.

What meteoric events happened in the Ninth Millenium? The tally is the Morro de Cuero Crater in Argentina, the Curtis Lake Crater near Palmer in Alaska, the Brushy Creek Impact Crater in St Helena Parish in Louisiana and the Parry Sound Impact Crater in Ontario in Canada. Not many really when compared to other millenia.

Then in the Eight Millenium everything exploded. For the period during the Eighth Millenium there were numerous impact events. Far more impact events than you would ever expect. The Rio Cuarto Crater Field in Cordoba Province in Argentina, the Joey Crater in the Southern Ocean south of Australia, the Kangaroo Crater in the Southern Ocean south of Australia, the Flinders Impact Crater in Bass Strait, the Hickman crater in Western Australia, tektites found in Victoria produced by an impact southeast of Tasmania, Otz Valley in Austria , the Merewether Crater in Labrador, the Hudson Bay Crater, the El Fayum or El Fayoum Depression in Egypt, the Tsoorikmae Impact Crater in Estonia, the Cabrerolles Crater in France, the Cape York Meteorite in Greenland, the Kukla Crater in the Southern Ocean,the Togyz Crater and the the Chelkar-Aralskaya Crater in Kazakhstan, the Baba Yaga Crater in North Korea, the Frombork and Morasko craters in Poland and Meteor Strike Crater in South Africa.

In Russia there was the Lezhninskoe Lake, the Vanelahti Crater, the Chukhlomskoye Lake, the Kostromskiye Razlivy, the Krugloe Lake, the Zabore Ozero, the Beloe-Bordukovskoe Ozero, the Chernoye-Bordukovskoe Ozero, the Eryomkovo Lake, Lake Karpovskoe, Lake Lemeshenskoe, the Muromskoe Ozero, the Nerskoye Ozero, the Orekhovo-Zuevskiye Crater, the Sherninskiye Ozero, the Smerdyachee Lake, the the Svyatoye Ozero, the Vvedenskiye Ozero, the Yarmoly Ozero, the Zolotaya Veshka Ozero, the Svetloyar Ozero, the Borovoye Ozero, the Chernogolvka 1 and Chernogolovka 2 craters, Lake Lukovoye, the Orlevo Ozero, the Svetloe Ozero, the Ozerskoe crater, the Izyar Lake, the Kuzmayar Lake, the Nestiar Ozero, the Tigoda Ozero, the Valday Crater, the Vershinskoye Crater, the Virovno Ozero, the Zamoshskoy Ozero, the Beloye-Tverskoye Ozero, the Shlepetskoyo Ozero, the Sutokovyi Crater, the Tarusovskiye Ozero, the Chyornye Crater, the Svetetskiye Crater, Urema Lake, the Srednikovo Crater, the Yulovskiy Ozero, the Otno Ozero, the Yazhino, Timokhino and Osinovik Craters, the Siglinitsy Ozero and the Talets Ozero.

Then it quietened down again.

Remember the damage that could be caused by a single impact around one mile across. Let me repeat it for one single impact.
What happens when we have an impact of this size? This impact would have propelled seismic waves across the face of the earth creating scores of earthquakes which themselves would have generated tsunamis hundreds of metres high. The sky would have become red hot as the atmosphere would fill with dust and at the same time the tops of the oceans would have boiled. The impact would have vapourized rocks which would have gone up into the atmosphere before condensing in to liquid droplets that solidified and fell back to the surface. This is as well as meteoric errata falling as well. This would be like the book of Revelations come to Earth.What was happening on Earth at this time. More than you may think as well.

What is the meteorite tally for the Seventh Millenium? The Lake Hamilton Crater on the Eyre Peninsula of South Australia, the Weepra Park crater near Elliston on the Eyre Peninsula of South Australia, the Big Basin Crater in Clark County in Kansas and the Malha Crater Lake near Kutum in Sudan.

The rest would be history but all of these impacts, especially in the Eight Millenium, have never been told to us or taught to us. The real nature of our position in the cosmos is nothing like the sanitized version that we have been left to believe.

Now we have the truth of what happened at the end of the last Ice Age!

This is a mixture of 9,000 BC and 8,000 BC impact events in Canada and Alaska. Sithylemenkat, Amundsen Gulf, Hudson Strait, Ungaca Bay and Merewether Craters are almost on a great curve of the earth. The Cape York crater, Baffin Bay, Hudson Strait and Hudson Bay craters also form a line. Or was it a very wide meteor stream over Canada and Alaska? Below you can see the Bermuda Crater which would have vapourized any Atlantic landmass. Was it the initial impact and the others followed from the southeast to the northwest?

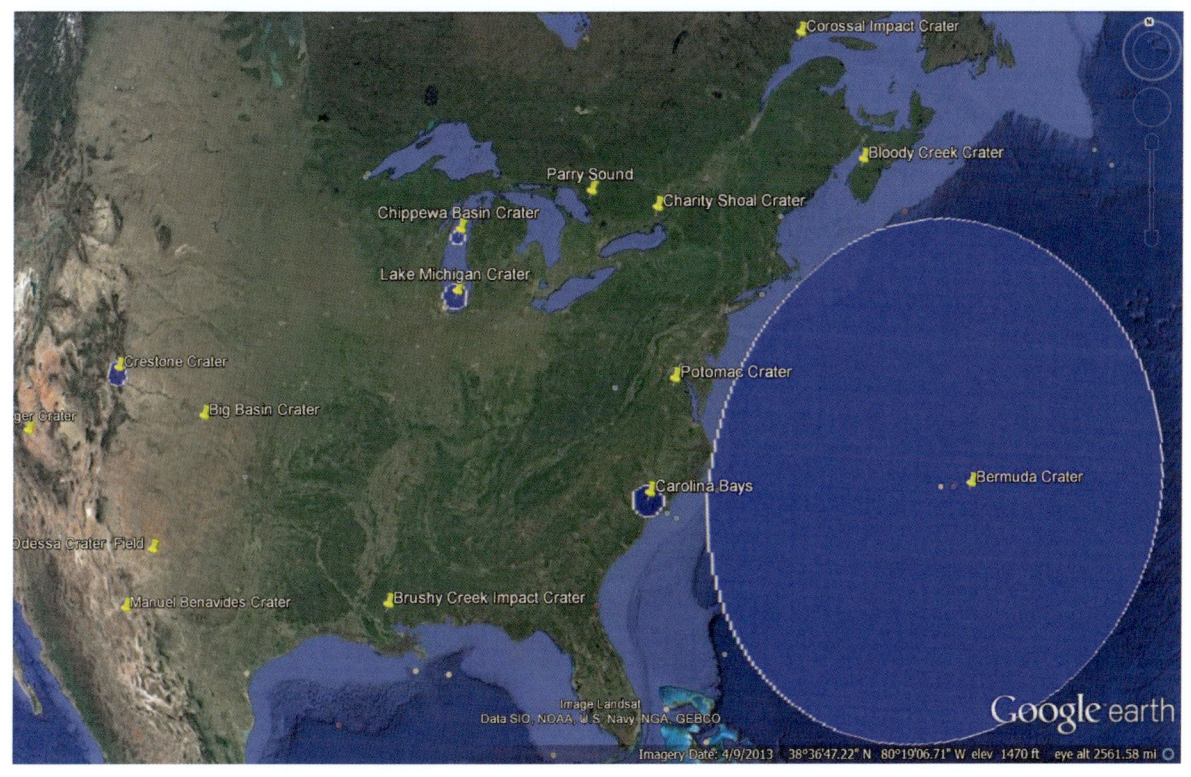

Remember that these events are occurring within a three thousand year period. They would certainly be memorable and as you can see prolific! It is like an impactoid storm.

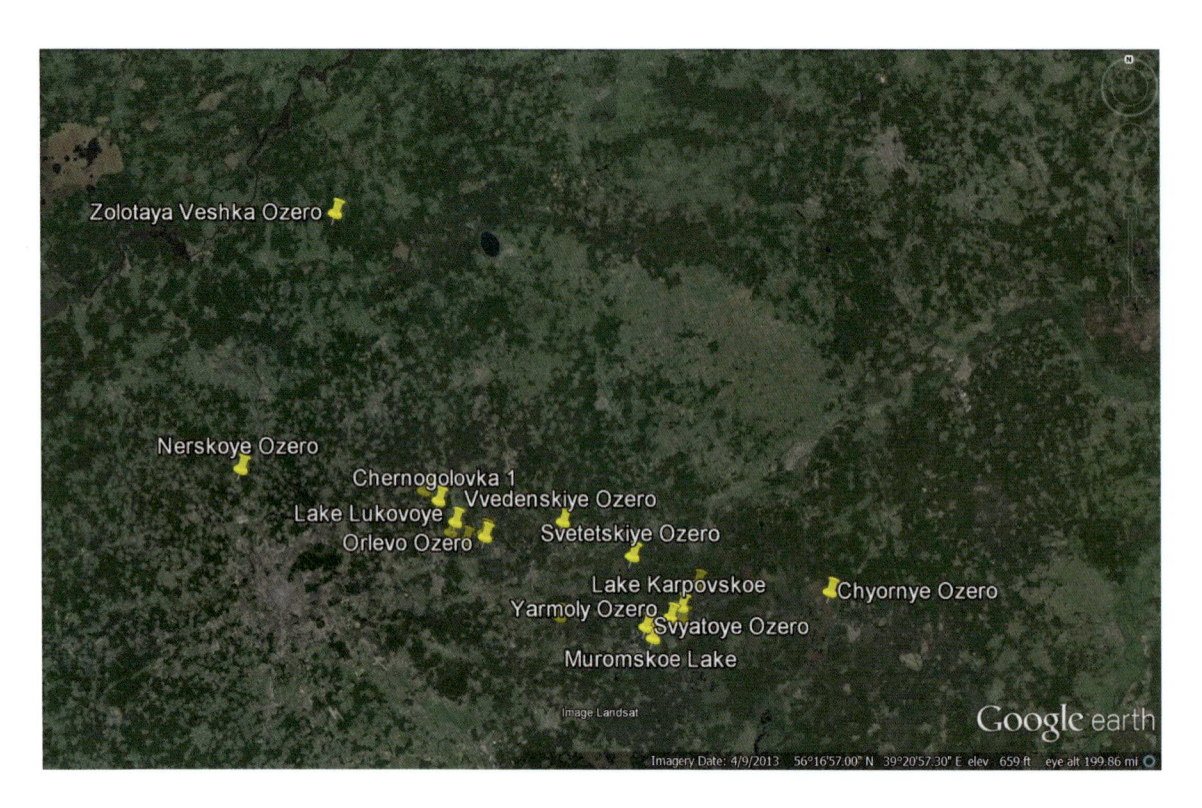

Google Earth satellite image showing impact crater locations near Moscow, Russia, with yellow pins marking: Zolotaya Veshka Ozero, Nerskoye Ozero, Chernogolovka 1, Lake Lukovoye, Orlevo Ozero, Vvedenskiye Ozero, Svetetskiye Ozero, Lake Karpovskoe, Yarmoly Ozero, Svyatoye Ozero, Muromskoe Lake, Chyornye Ozero.

What is the coincidental value of two massive meteorite showers hitting the same area in Russia in three thousand years? Nil would probably be the best answer. What would be the statistical probability of it being a single meteorite shower resulting in a mass of impacts at the same time? This would be a certainty.

This is not the case with the Bermuda Crater though which is the best indicator of there having been two impact events. The first in the Eleventh Millenium and the second in the Eighth Millenium. This might not be the case for the Russian impacts though. The Russian impacts appear to have occurred around the same time.

How many impact craters do you want? These are the ten thousand years old impact craters near Moscow in Russia. What did the sky look like in this period as they came crashing down? Were they parts of a single rubble asteroid that broke up as it hit our atmosphere? Rubble asteroids? A rubble asteroid is not a single monolith but a mass of numerous pieces of rock that have coalesced under the influence of a gravitational field. These rubble piles or rubble asteroids form when an asteroid or moon is smashed to pieces by an impact. The shattered pieces fall back together due to self-gravitation. These asteroids have low densities and are prone to explode in the atmosphere whilst descending. Many asteroids and even several planetary moons are now believed to be orbiting rubble piles.

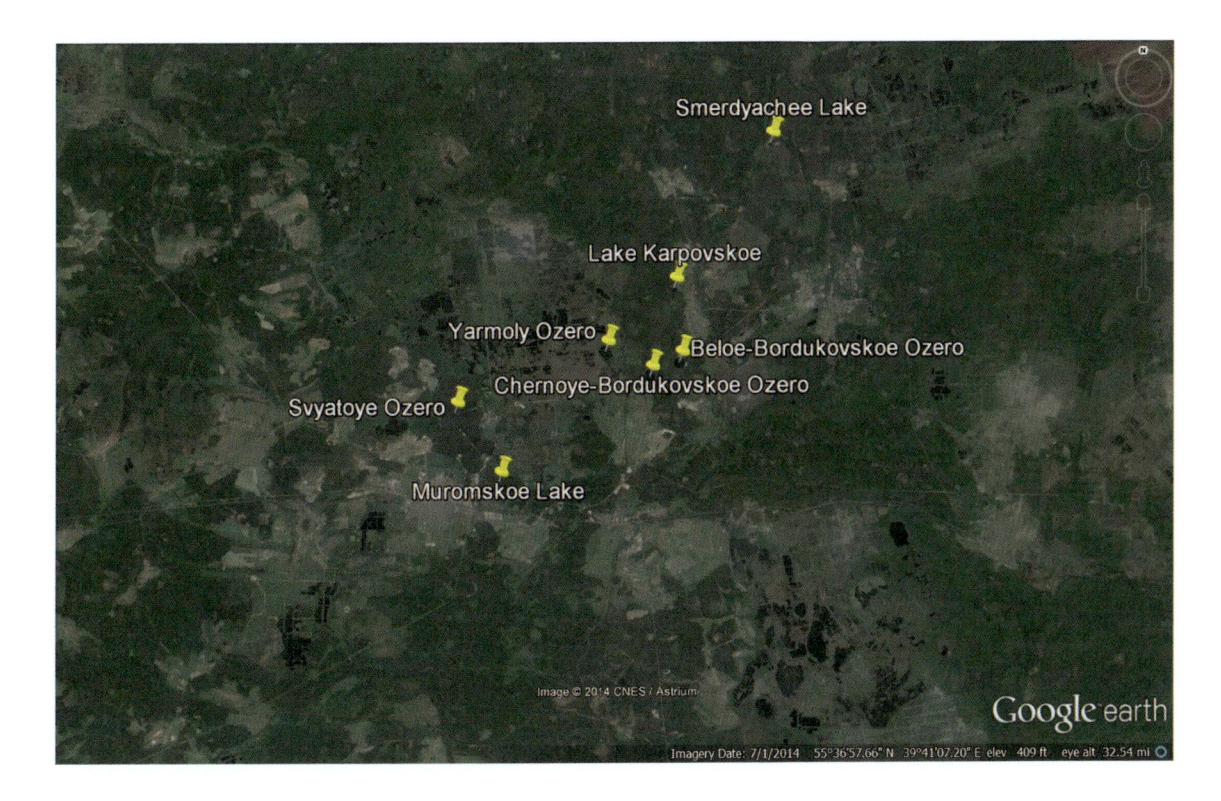

We head further towards the right-hand portion of the crater field.

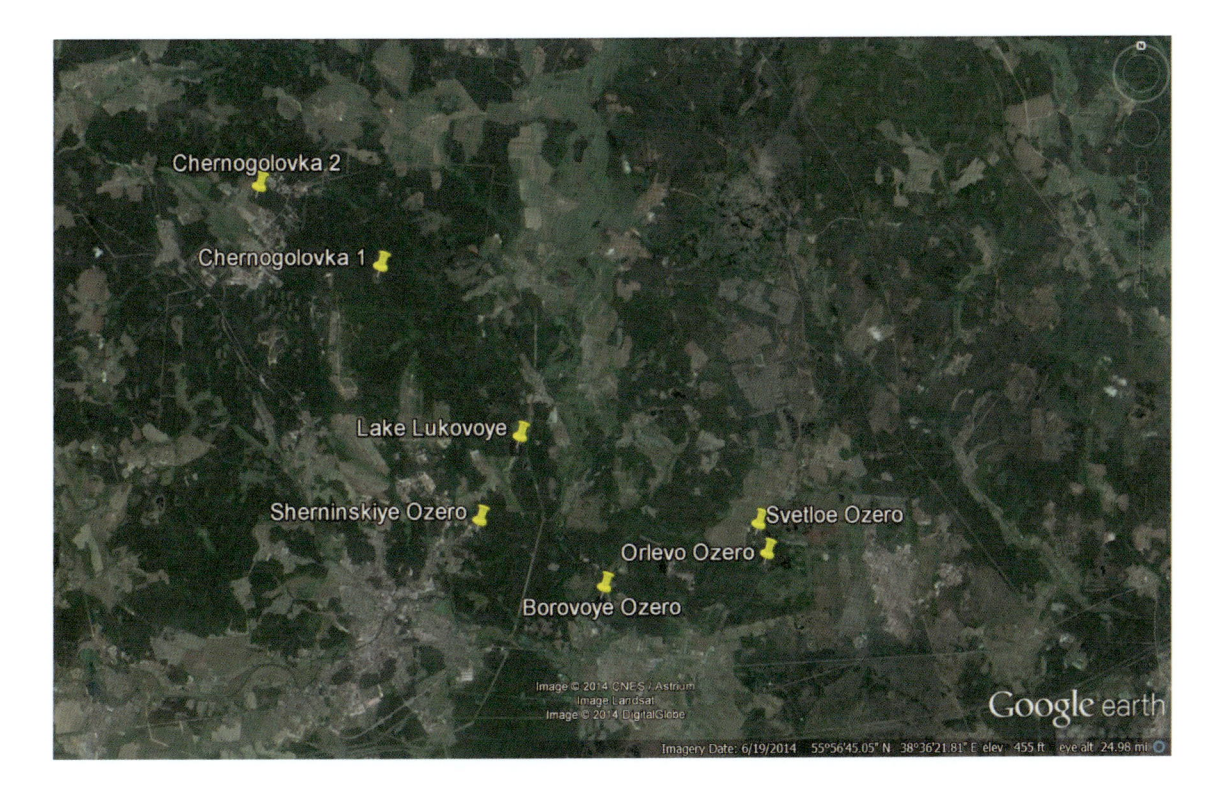

Russian impact craters in the left-hand portion of the crater field.

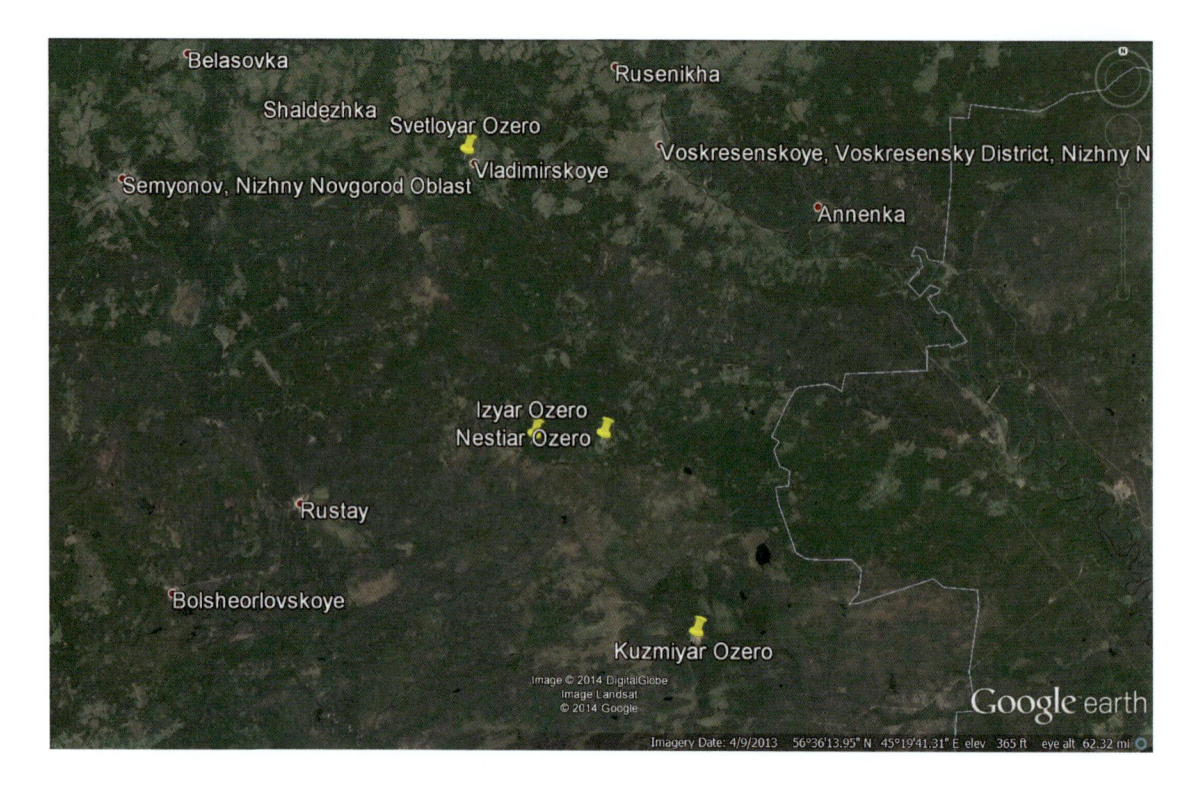

Impacts in Nizhegorodskaya Oblast in Russia to the east of Moscow in this same period.

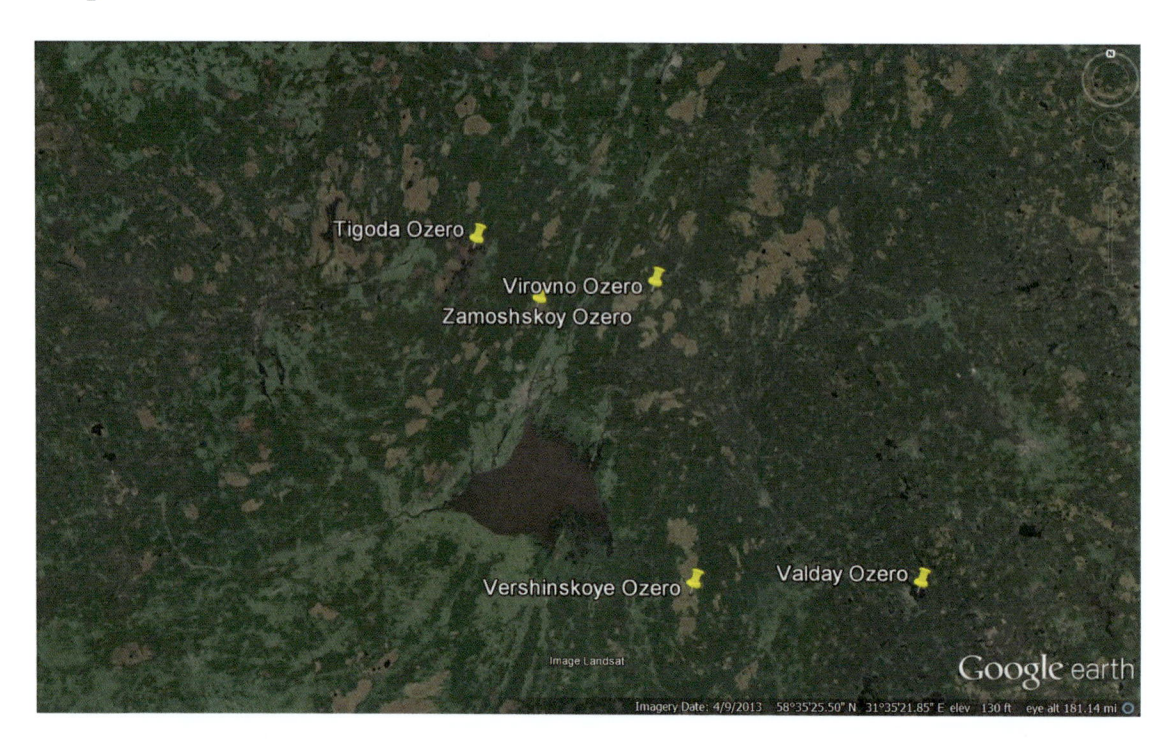

Impacts in Novgorodskaya Oblast in Russia around 8,000 BC.

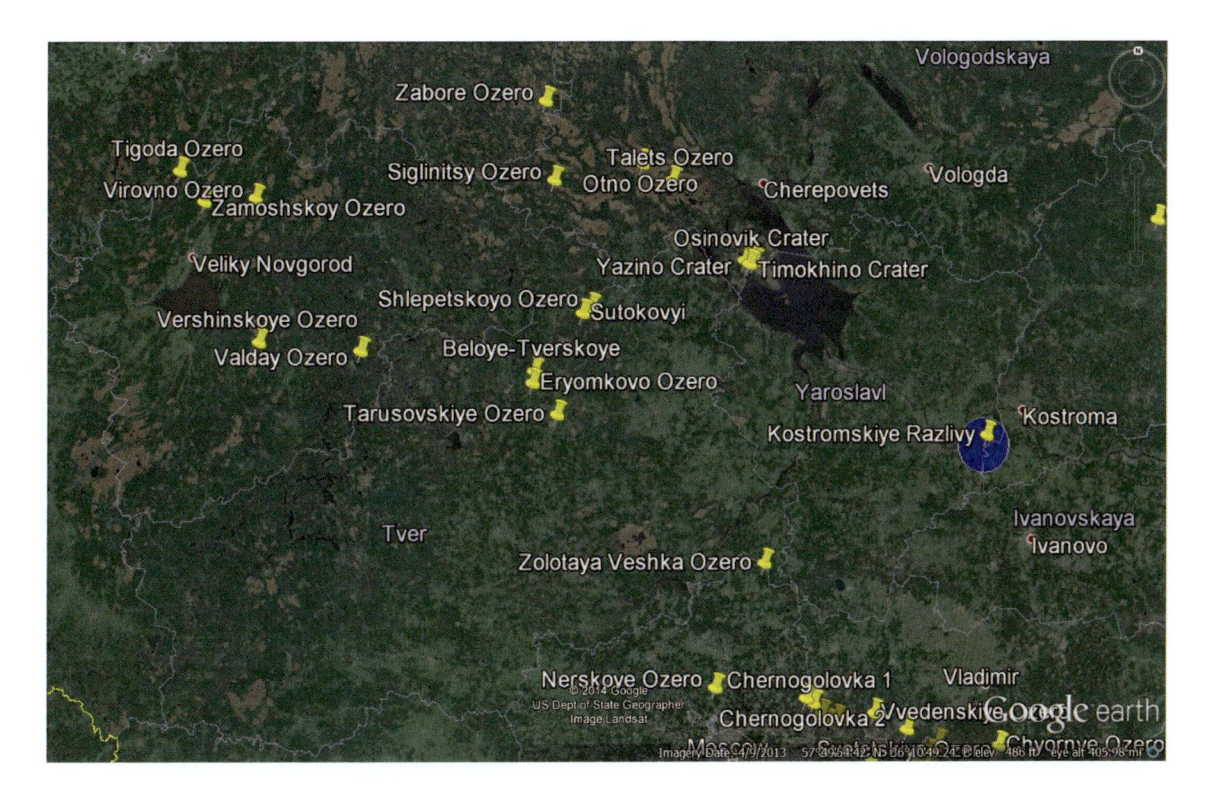

Impact craters in Tverskaya Oblast, Vladimirskaya Oblast and Vologdaya Oblast in Russia.

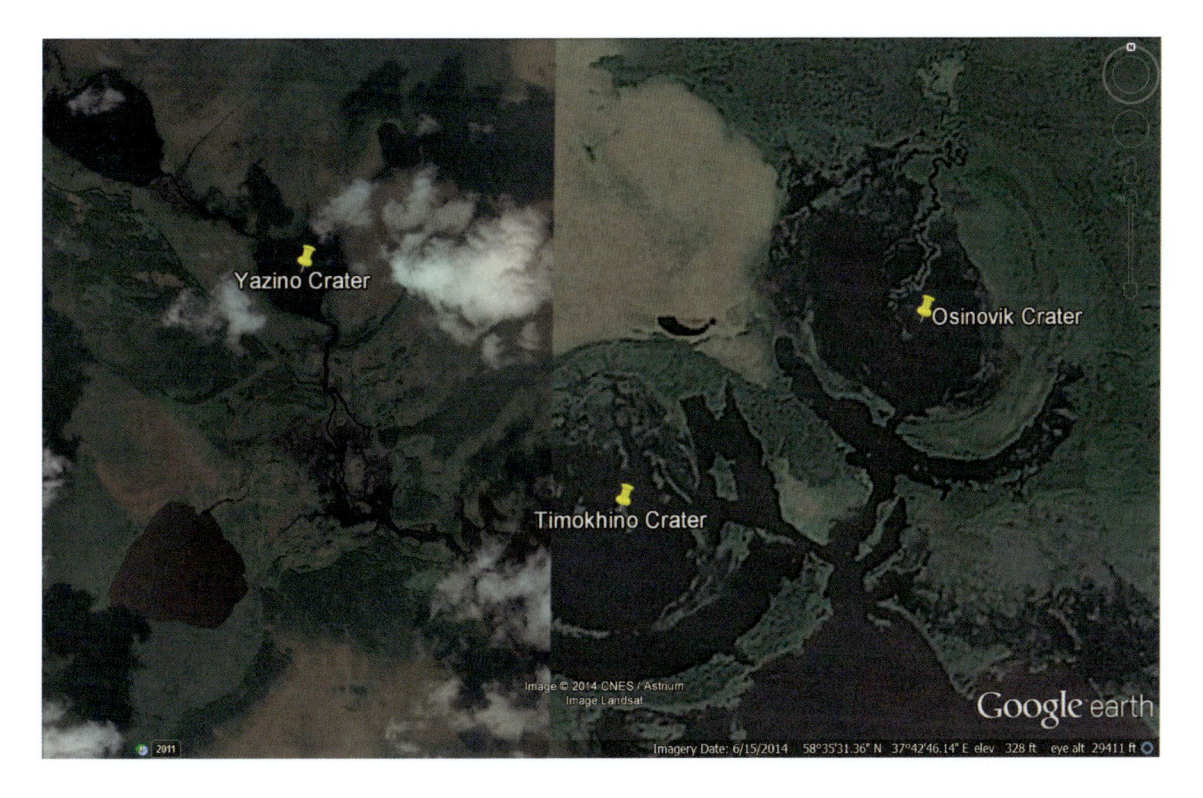

Yazhino, Timokhino and Osinovik Craters in Vologdaya Oblast. How does normal science explain this clustering of craters? Especially during the largest meteorite storm to ever hit the Earth! No wonder the ancients commemorated this with their obscession with astronomy.

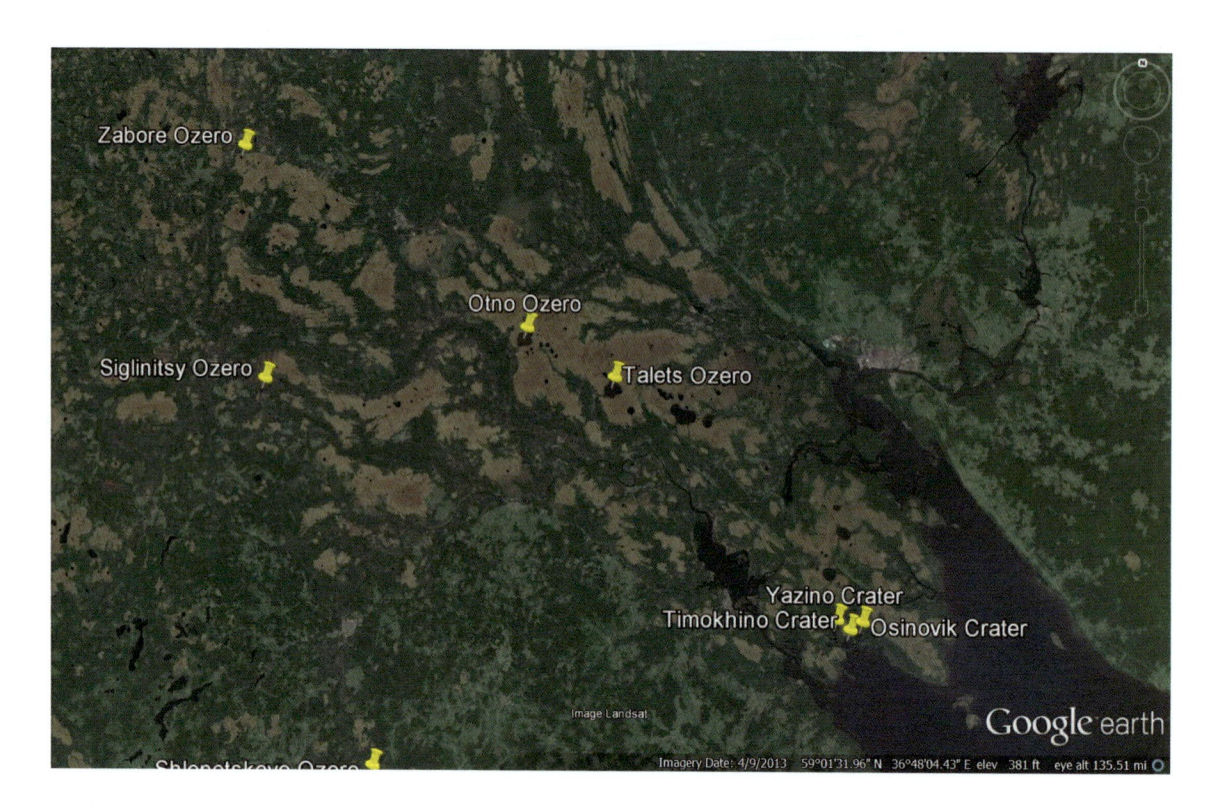

Was this a northwest to southeastern impact stream or the opposite?

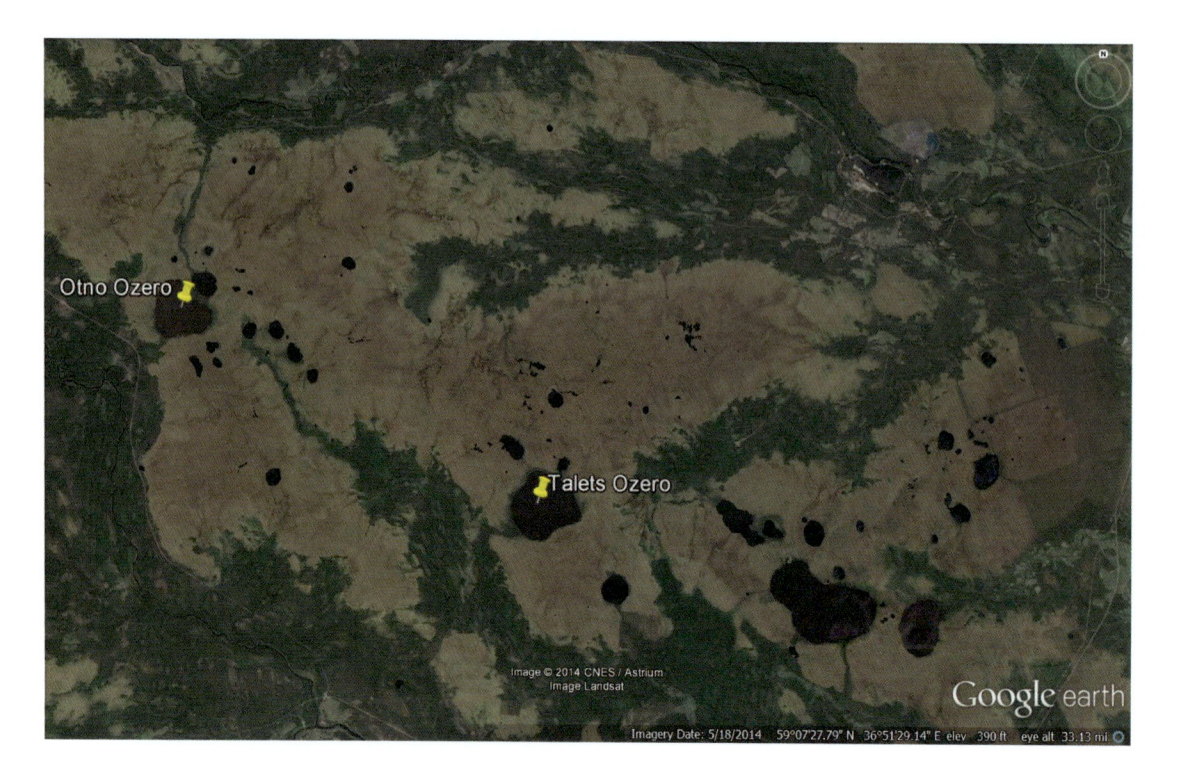

Are the circular and elliptical lakes in this photo also impact craters? They even overlap the same as the Carolina Bays in North America.

Allowing for geological variance can I ask if the Russian impact craters that are dated to 11,000 BC and 10,900 BC might even be from this same period as dating is very confused in this period. This would make the impactoid showers in this period the most spectacular ever seen by the human race.

This map gives you an indication of how widespread the meteorite shower or storm was. From Siberia to France. And these are only the known ones. How many more are under the seas or lost in deserts of mountain ranges?

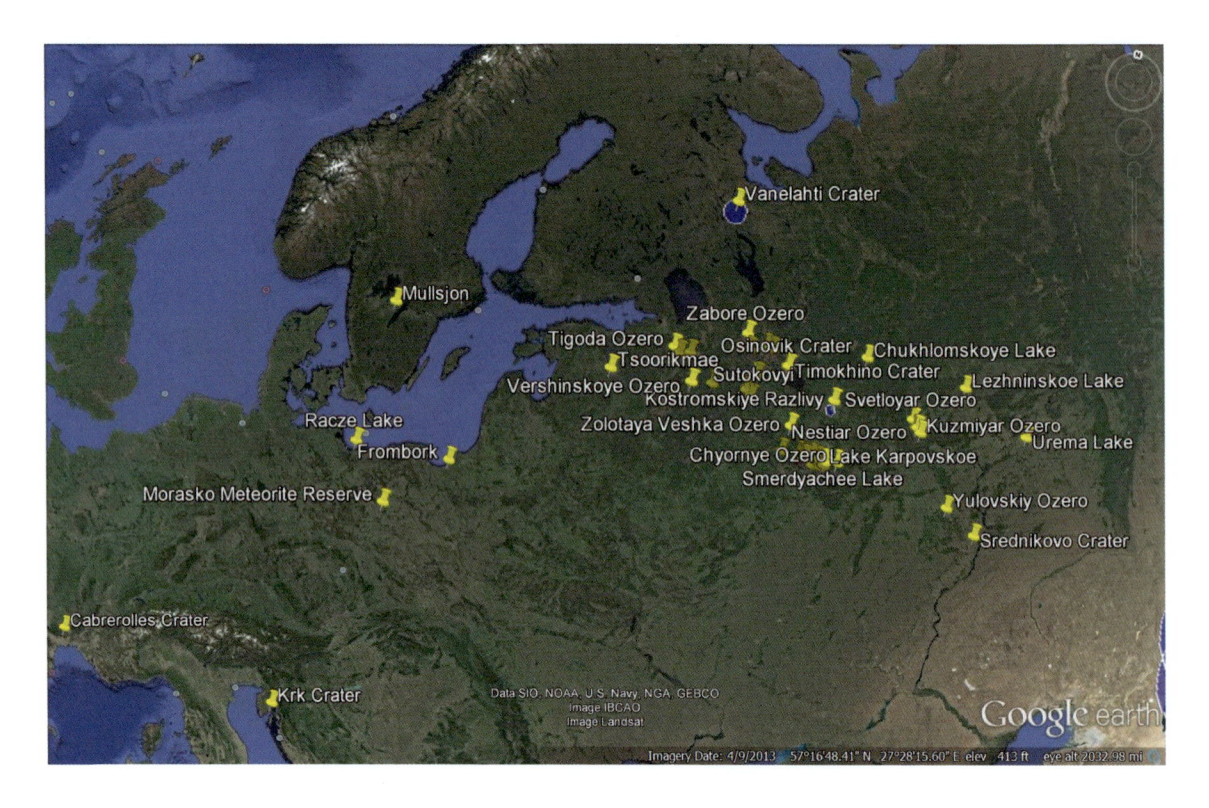

Do we really need words here now? What sort of impact stream was this? Other than an enormous one!

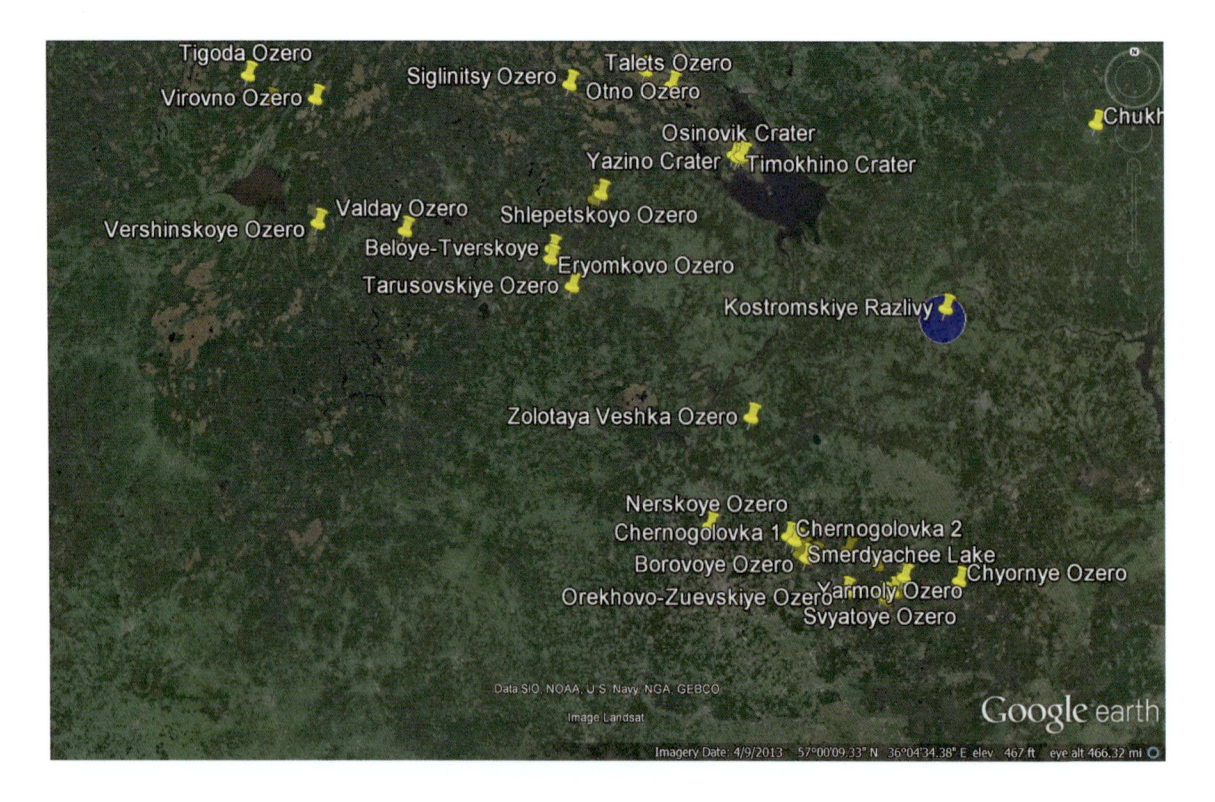

Which way was the impact stream heading?

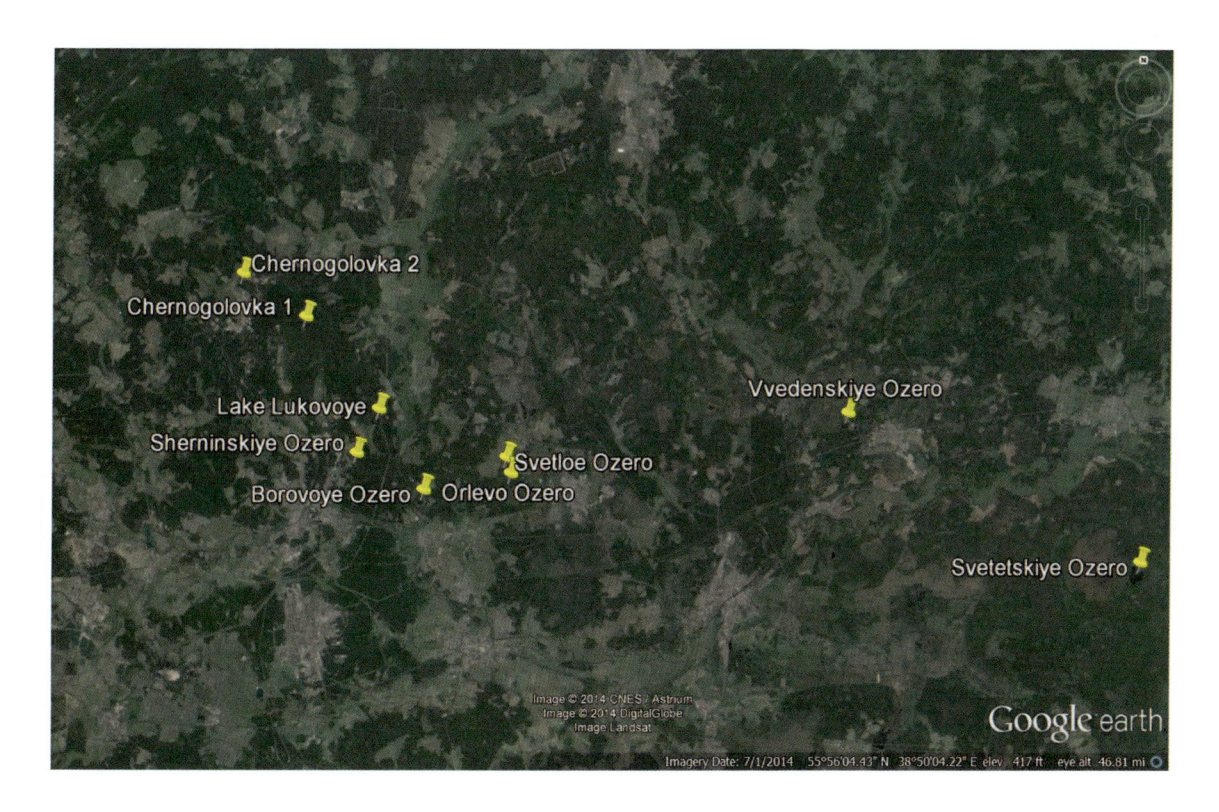

A close up of more Russian impact craters.

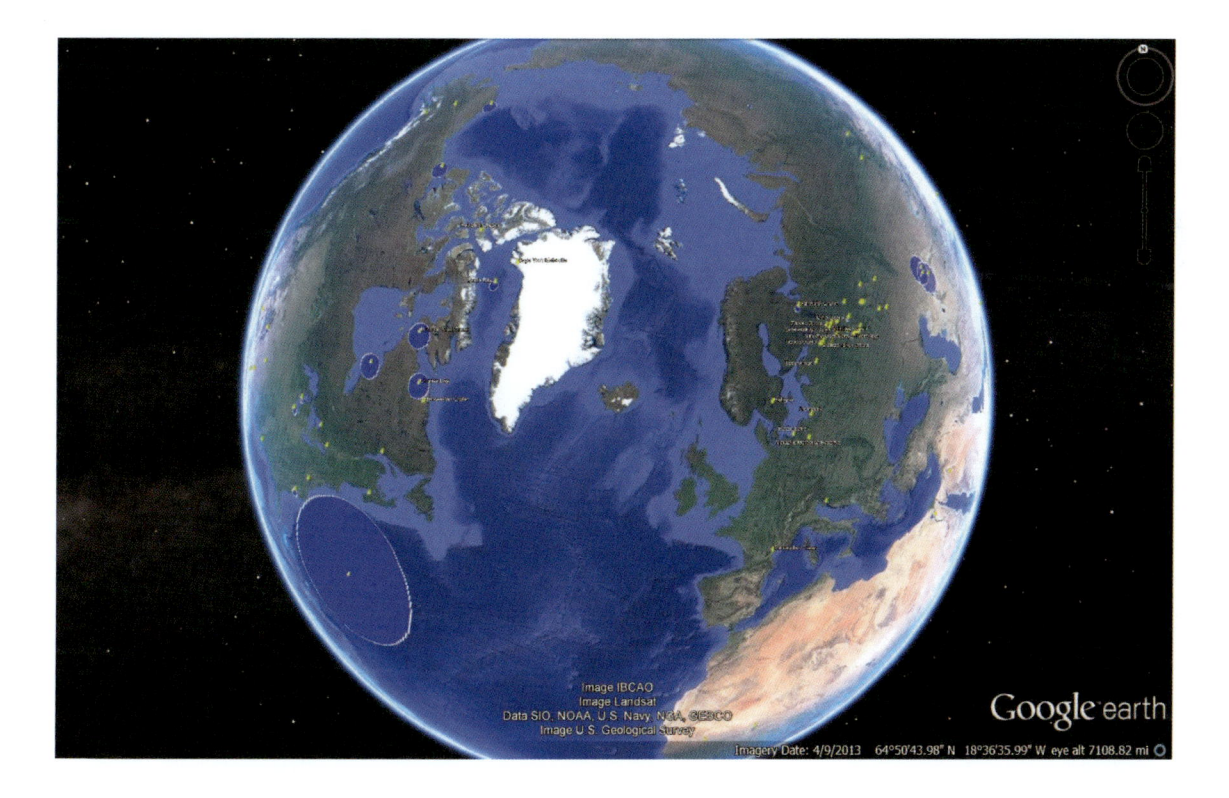

The big blue area to the left is the Bermuda Crater at the bottom of an impact stream across Canada and to Alaska. To the right is the Russian mega-strewn field that appears as if a bucket of gravel was thrown at the earth covering a very large elliptical area. Allowing for the nature of newly discovered rubble asteroids the cosmic body that struck the right-hand part of the above map seems to have fragmented. Did Earth have an impact with a binary pair of cosmic bodies in this period or was it a loosely bonded together peanut-shaped rubble asteroid?

    Scientists were originally puzzled by the densities of some asteroids as they appeared less dense than the meteorites that composed them and that in fact had come from them and were in fact pieces of asteroids.
Large interior voids are possible in these asteroids because of their low gravity. Many of them may even have what appears to be a fine and solid-looking regolith or covering on the outside but the asteroid's gravity is so weak that that friction between fragments dominates and prevents small pieces from falling inwards and filling up the voids.
    As an example the asteroid Itokawa appears to be two contact binaries, that is two asteroid bodies touching, with or without rubble filling the boundary between them. This is peanut-shaped the same as some recently observed comets. How well held together are comets then? Phobos, the largest of the two moons of Mars, is now considered to be a rubble asteroid.

You tell me what happened here then? The blue shaded areas indicate the crater sizes of the impacts? This seems to be a row of them in diminishing sizes. This is the history of the Earth as we have never been taught it!

A history of massive impacts of cosmic bodies that have shattered the peace and calm of the planet and at times have led to the almost complete extinction of life on our planet, and as we have seen the possible introduction of life as well. We have found that there were no Ice Ages as such but misidentified Polar ice masses relative to where the Poles would have been. We do not live on a gentle planet with subtle million year old changes but a planet subject to massive and sudden change. These changes caused the extinction of many of the creatures that inhabited the earth. And these changes almost caused the extinction of man, possibly several times.

This is a history of the Earth that you have never dreamt of.
This is a history of the Earth that you have never been taught.
This is a history of the Earth that you never thought possible.
This is a history of the greatest cataclysm in recent times.
This is what happened at the end of the last Ice Age and changed the face of the Earth and the destiny of its life forms forever.
This is what you have never been taught in your schools and universities at all!

George Mitrovic
Picton 2018

Bibliography.

"Gods of the New Millenium" Alan F. Alford. New English Library. Hodder and Stoughton. London.1998.

"Cataclysm" D. S. Allan and J.B. Delair. Bear & Company. Rocheste, Vermont. 1997.

"Ancient Traces-Mysteries in Ancient and Early History". Michael Baigent. Penguin Books. London. 1999.

"Ancient Mysteries-A History Through Evolution and Magic". Michael Baigent. Penguin Books. London. 1998.

"Exodus to Arthur. Catastrophic Encounters with Comets." Mike Baillie. B. T. Batsford Ltd. London. 2000.

"Lost City of Stone". Bill S. Ballinger. Simon and Schuster. New York. 1978.

"Extraterrestrial Visitations from Prehistoric times to the Present". Jacques Bergier. New American Library. New York. 1974.

"The Dragon's Triangle." Charles Berlitz. Guild Publishing. Glasgow. 1989.

"Atlantis-The Lost Continent Revealed". Charles Berlitz. Fontana Paperbacks. London. 1985.

"The Mystery of Atlantis". Charles Berlitz. Panther Books. England. 1977.

"Mysteries from Forgotten Worlds". Charles Berlitz. Corgi Books. Great Britain. 1976.

"Unsolved Mysteries of the Past". Otto O. Binder. Tower Publications. New York. 1970.

"Prehistoric Man". Robert J. Braidwood. William Morrow and Company.New York. 1967.

"Lost Atlantis" James Bramwell. Freeway Press. New York. 1973.

"Weird America: A Guide to Places of Mystery in the United States". Jim Brandon. Dutton Paperbacks. New York. 1978.

"The Ancient Earthworks and Temples of the American Indians". Lindesay Brine. Oracle Publishing. England. 1996.

"Comets. Speculation and Discovery". Nigel Calder. Dover Publications, Inc. New York. 1994.

"Lost Worlds. Scientific Secrets of the Ancients". Robert Charroux. Fontana/Collins Press. Glasgow. 1974.

"Masters of the World". Robert Charroux. Sphere Books. Great Britain. 1979.

"One Hundred Thousand Years of Man's Unknown History". Robert Charroux. Berkeley Publishing. New York. 1971.

"The Mysterious Past". Robert Charroux. Futura Publications. London. 1974.

"Our Ancestors Came From Outer Space". Maurice Chatelain. Pan Books. London. 1979.

"Lost Cities of Ancient Lemuria and the Pacific". David Hatcher Childress. Adventures Unlimited Press. Stelle, Illinois. 1988.

"Lost Cities of Atlantis, Ancient Europe and the Mediterranean". David Hatcher Childress. Adventures Unlimited Press. Stelle, Illinois. 1996.

"Lost Cities of China, Central Asia and India". David Hatcher Childress. Adventures Unlimited Press. Stelle, Illinois. 1991.

"Lost Cities and Ancient Mysteries of South America". David Hatcher Childress. Adventures Unlimited Press. Stelle, Illinois. 1989.

"Lost Cities and Ancient Mysteries of Africa and Arabia". David Hatcher Childress. Adventures Unlimited Press. Stelle, Illinois. 1997.

"Lost Cities of North and Central America". David Hatcher Childress. Adventures Unlimited Press. Stelle, Illinois. 1993.

"Watermark". Joseph Christy-Vitale. Paraview Pocket Books. New York. 2004.

"Catastrophobia". Barbara Hand Clow. Bear & Company. Rochester, Vermont. 2001.

"The Ancient Visitors". Daniel Cohen. Doubleday and Co, Inc, Garden City, New York. 1976.

"The Fallen Sky". Christopher Cokinos. Jermey P. Tarcher/Penguin. New York. 2009.

"Gods of Eden". Andrew Collins. Headline Book Publishing. London. 1998.

"The Cygnus Mystery". Andrew Collins. Watkins Publishing. London. 2006.

"Gateway to Atlantis". Andrew Collins. Carroll and Graf Publishers, Inc. New York. 2000.

"The New Pyramid Age". Philip Coppens. O Books. Winchester. United Kingdom. 2007.

"Architects of Eternity." Richard Corfield. Headline Book Publishing. 2001.

"Strange Artifacts 1-A Sourcebook on Ancient Man". William R. Corliss. The Sourcebook Project. Glen Arm, Maryland. 1975.

"Strange Artifacts 2-A Sourcebook on Ancient Man". William R. Corliss. The Sourcebook Project. Glen Arm, Maryland. 1976.

"Ancient Man: A Handbook of Puzzling Artifacts". William R. Corliss. The Sourcebook Project. Glen Arm, Maryland.1976.

"Carolina Bays, Mima Mounds, Submarine Canyons and other Topographical Phenomena." William R. Corliss. The Sourcebook Project. Glen Arm, Maryland. 1988.

"Unknown Earth: A Handbook of Geological Enigmas." William R. Corliss. The Sourcebook Project. Glen Arm, Maryland. 1980.

"Anomalies in Geology: Physical, Chemical, Biological." William R. Corliss. The Sourcebook Project. Glen Arm, Maryland. 1989.

"Neglected Geological Anomalies." William R. Corliss. The Sourcebook Project. Glen Arm, Maryland. 1990.

"Inner Earth: A Search for Anomalies". William R. Corliss. The Sourcebook Project. Glen Arm, Maryland. 1991.

"Forbidden Archeology". Michael A. Cremo and Richard L. Thompson. Bhaktivedanta Book Publishing. 1998.

"Earth's Most Challenging Mysteries". Reginald Daly. The Craig Press. United States of America. 1972.

"Voyagers to the New World". Nigel Davies. William Morrow and Co. Inc. New York. 1979.

"The Lost Tribes From Outer Space". Mark Dem. Corgi Books. London. 1977.

"Mysterious Ancient America". Paul Devereux. Vega. London. 2002.

"Atlantis-The Antedeluvian World". Ignatius Donnelly. Steiner Books. New York. 1975.

"The Destruction of Atlantis. Ragnarok: The Age of Fire and Gravel". Ignatius Donnelly. Steiner Books. Blauvelt, New York. 1971.

"Strange World". Frank Edwards. Lyle Sturat. Secauscas, New Jersey. 1974.

"The Long Summer". Brian Fagan. Basic Books. New York. 2004.

"The Cycle of Cosmic Catastrophes". Richard Firestone, Allen West and Simon Warwick-Smith. Bear and Company. Rochester. Vermont. 2006.

"When the Sky Fell-In Search of Atlantis". Rand and Rose Flem-Ath. Weidenfeld and Nicholson. London. 1995.

"Fingerprints of the Gods-A Quest for the Beginning of the End". Graham Hancock. Mandarin. London. 1995.

"The Mars Mystery". Graham Hancock, Robert Bauval and John Grigsby. Michael Joseph Ltd. Penguin Books. Ringwood. Victoria. Australia. 1998.

"Path of the Pole". Charles H. Hapgood. Adventures Unlimited Press. Kempton, Illinois. 1999.

"Beyond Stonehenge". Gerald S. Hawkins. Arrow Books. London. 1973.

"Fatu-Hiva. Back to Nature". Thor Heyerdahl. George Allen and Unwin and Associates. London. 1974.

"Aku-Aku. The Secret of Easter Island". Thor Heyerdahl. George Allen and Unwin Limited. 1958.

"The Tigris Expedition". Thor Heyerdahl. George Allen and Unwin. London. 1980.

"The Kon-Tiki Expedition". Thor Heyerdahl. George Allen and Unwin Ltd. London. 1951.

"The Maldive Mystery". Thor Heyerdahl. Unwin Paperbacks. London. 1988.

"The Dragon and the Disc-An Investigation of the Totally Fantastic". F. W. Holiday. W. W. Norton. New York. 1973.

"Sons of the Sun". Marcel F. Homet. Neville Spearman. London. 1963.

"In Quest of the White God". Pierre Honore. Futura Publications. London. 1975.

"Atlantis: Myth or Reality?" Murry Hope. Arkana Penguin Books. London. 1991.

"Evolution from Space". Fred Hoyle and N. C. Wickramasinghe. J. M. Dent and Sons. London. 1981.

"Diseases from Space". Fred Hoyle and N. C. Wickramasinghe. Harper and Row. New York. 1979.

"Flood Stories from Around the World". Mark Isaak. Mirrored from http:home.earthlink.net/~misaak/floods.htm

"Survivors of Atlantis" Frank Joseph. Bear and Company. Rochester Vermont. 2004.

"The Atlantis Encyclopedia" Frank Joseph. New Page Books. Franklin Lakes New Jersey. 2005.

"Atlantis and Other Lost Worlds". Frank Joseph. Arcturus Publishing. London. 2008.

"Impact Geology". Allan O. Kelly. Typesetting by Headline Advertising, Encinitas, California. 1985.

"Target: Earth. The Role of Large Meteors In Earth Science". Allan. O. Kelly and Frank Dachille. Pensacola Engraving Co, Inc. Pensacola, Florida. 1953.

"Uriel's Machine". Christopher Knight and Robert Lomas. Arrow Books. London. 2000.

"Civilization One". Christopher Knight and Alan Butler. Watkins Publishing. London. 2004.

"Not of this World-Mysteries of Our Ancient Past". Peter Kolosimo. Souvenir Press. London. 1970.

"Timeless Earth". Peter Kolosimo. Sphere Books. London. 1974.

"Spaceships in Prehistory". Peter Kolosimo. University Books. Secaucus, NJ. 1976.

"In Search of Lost Civilizations". Alan Landsburg. Corgi Books. London. 1977.

"In Search of Ancient Mysteries". A & S. Landsburg. Corgi Books. London. 1974.

"Earth Under Fire". Paul A. La Violette, PH.D. Bear and Company. Rochester Vermont. 2005.

"Temple of the Stars". Brinsley le Poer Trench. Fontana/Collins Books. London. 1973.

"Comets. Creators and Destroyers" . David H. Levy. Toughstone. New York. 1998.

"Rain of Iron and Ice". John S. Lewis. Helix Books. Addison-Wesley Publishing Company. 1996.

"Myths of Pre-Columbian America". Donald. A. Mackenzie. Gresham Publishing. London. No date. Late Nineteenth-Early Twentieth Century.

"Myths and Traditions of the South Sea Islands". Donald A Mackenzie. Gresham Publishing Company. London.1930.

"Myths and Legends of China and Japan". Donald A. Mackenzie. Bracken Books. London. 1992.

"Before the Pharaohs: Egypt's Mysterious Prehistory". Edward F. Malkowski. Bear and Company. Rochester. Vermont.

"The Lost Empire of Atlantis". Gavin Menzies. Swordfish. Orion Publishing. London. 2011.

"Flood Shock. The Drowning of Planet Earth" Antony Milne. Alan Sutton Publishing. Gloucester.

"Atlan Revisited- The War of the Gods". George Mitrovic. Vantage Press, New York. 1985.

"When the Sky Crashed Down". George Mitrovic. Createspace Independent Publishing Platform. 1582. 2014.

"Colony Earth". Richard E. Mooney. Fawcett Publications. Grenwich,Connecticut. 1974.

"The Secret of Atlantis." Otto Muck. William Collins Sons & Co Ltd. London. 1978.

"Secrets of the Lost Races". Rene Noorbergen (researched for by J. B. Jochmans). Bobbs: Merrill. New York. 1977.

"The Eternal Man". Louis Pauwels. Avon Books. New York. 1972.

"Impossible Possibilities". Louis Pauwels. Stein and Day. New York. 1971.

"The End of Eden- The Comet That Changed Civilization" Graham Phillips. Bear and Company. Rochester. Vermont. 2007

"The Templars and the Ark of the Covenant". Graham Phillips. Bear and Company. Rochester, Vermont. 2004.

"Sungods in Exile-The Secrets of the Dzopa of Tibet". Karyl Robin-Evans. Neville Spearman. Sudbury. England. 1978.

"Disturbing the Solar System". Alan E Rubin. Princeton University Press. Princeton. 2002.

"Noah's Flood" William Ryan and Walter Pitman. Simon and Schuster. New York. 1998.

"Comet". Carl Sagan and Ann Druyan. Pocket Books. New York. 1985.

"Investigating the Unexplained". Ivan T. Sanderson. Prentise-Hall. Englewood Cliffs. New Jersey. 1972.

"Voyages of the Pyramid Builders". Robert M. Schoch, Ph. D, with Robert Aquinas McNally. Jeremy P. Tarcher/ Penguin Books. New York. 2004.

"Comets. Vagabonds of Space". David A Seargent. Doubleday. Sydney and Auckland. 1982.

"The Incredible Discovery of Noah's Ark." Charles E. Sellier and David W. Balsiger. Dell Publishing, New York. 1995.

"The Neandertal Enigma". James Shreeve. William Morrow and Co, Inc. New York. 1996.

"Lost Cities of the Ancients-Unearthed". Warren Smith. Zebra Books. New York. 1976.

"Atlantis of the North". Jurgen Spanuth. Sidgwick and Jackson. London. 1979.

"Lost Continents-The Atlantis Theme in History, Science and Literature". L. Sprague de Camp. Dover Books. New York. 1970.

"Atlantis Discovered." Lewis Spence. Causeway Books. New York. 1974. The Book Tree. San Diego. California. 2002.

"The Problem of Lemuria". Lewis Spence.

"Rogue Asteroids and Doomsday Comets". Duncan Steele. John Wiley and Sons, Inc. Toronto. 1995.

"Target Earth". Duncan Steel. The Readers Digest Association Inc. Pleasantville, New York/Monteal. 2000.

"Atlantis Rising". Brad Steiger. Sphere Books. London. 1977.

"Worlds Before Our Own." Brad Steiger. W. H. Allen. London. 1980.

"Cosmic Pinball". Carolyn Summers and Carlton Allen. McGraw Hill. NewYork. 2000.

"The Sirius Mystery". Robert K. G. Temple. Futura Publications. Great Britain. 1976.

"On the Shores of Endless Worlds". Andrew Tomas. Souvenir Press. London. 1974.

"Atlantis-From Legend to Discovery". Andrew Tomas. Robert Hale. London. 1972.

"We Are Not The First." Andrew Tomas. Souvenir Press. London. 1971.

"Mystery of the Ancients". Eric and Craig Umland. Panther Books. Great Britain. 1975.

"Worlds in Collision". Immanuel Velikovsky. Victor Gollancz Ltd. London, England. 1973.

"Ages in Chaos". Immanuel Velikovsky. Sidgwick and Jackson Limited. London. 1952.

"Earth in Upheaval". Immanuel Velikovsky. Abacus Books. Great Britain. 1974.

"Impact. The Threat of Comets and Asteroids". Gerrit. L. Verschur. Oxford University Press. New York. 1997.

"Lost Survivors of the Deluge". Gerd Von Hassler. Signet Books. New York. 1978.

"Comets and the Origin of Life". Janaki Wikramasinghe, Chandra Wickramasinghe and William Napier. World Scientific. Singapore. 2010.

"Secret Places of the Lion". George Hunt Williamson. Futura Publications. 1974.

"Alien Dawn- An Investigation into the Contact Experience". Colin Wilson. Virgin Publishing. London. 1998.

"From Atlantis to the Sphinx" Colin Wilson. Virgin Publishing. London. 2007.

"Atlantis and the Kingdom of the Neanderthals" Colin Wilson. Bear and Company. Rochester. Vermont. 2006.

"The Atlantis Blueprint". Colin Wilson and Rand Flem-Ath. Delta Trade Paperbacks. Dell Publishing. New York. 2002.

"Lost Outpost of Atlantis". Richard Wingate. Everest House Publishers. New York. 1980.

"Axis of the World". Igor Witkowski. Adventures Unlimited Press. Kempton, Illinois. 2008.

"Comets. A Chronological History of Observation, Science, Myth and Folklore." Donald K Yeomans. Wiley Science Editions. John Wiley and Sons, Inc. New York. 1991.

"The Flood from Heaven". Eberhard Zangger. William Morrow and Company, Inc. New York. 1992.

"The Stones of Atlantis". Dr. David Zink. W. H. Allen and Company. London. 1978.

Expert Database On Earth Impact Structures. gvk@sscc.ru

The Complete Catalog of the Earth's Impact structures of Anna Mikheeva, ICM&MG SB RAS. http://labmpg.sscc.ru/impact/index1.html

The SAO/NASA Astrophysics Data System. www.adsabs.harvard.edu/

Printed in Poland
by Amazon Fulfillment
Poland Sp. z o.o., Wrocław